AF480868

CLOUD TECHNOLOGIES WITH A FOCUS ON AI INTEGRATION AND AUTOMATION

Vidyasagar Vangala

Copyright © Vidyasagar Vangala
All Rights Reserved.

This book has been self-published with all reasonable efforts taken to make the material error-free by the author. No part of this book shall be used, reproduced in any manner whatsoever without written permission from the author, except in the case of brief quotations embodied in critical articles and reviews.

The Author of this book is solely responsible and liable for its content including but not limited to the views, representations, descriptions, statements, information, opinions and references ["Content"]. The Content of this book shall not constitute or be construed or deemed to reflect the opinion or expression of the Publisher or Editor. Neither the Publisher nor Editor endorse or approve the Content of this book or guarantee the reliability, accuracy or completeness of the Content published herein and do not make any representations or warranties of any kind, express or implied, including but not limited to the implied warranties of merchantability, fitness for a particular purpose. The Publisher and Editor shall not be liable whatsoever for any errors, omissions, whether such errors or omissions result from negligence, accident, or any other cause or claims for loss or damages of any kind, including without limitation, indirect or consequential loss or damage arising out of use, inability to use, or about the reliability, accuracy or sufficiency of the information contained in this book.

Made with ❤ on the Notion Press Platform
www.notionpress.com

PREFACE

Cloud computing and AI and automated technologies are creating new trends in the technological environment as the implementation of these technologies is accelerated in different industries. With organizations using digital technologies more and more, integration of these technologies has become fundamental for mass, flexibility, and analytics.

Cloud Technologies with a Focus on AI Integration and Automation examines this potent partnership, offering a practical reference for executives, data scientists and analysts, and tech business visionaries interested in cloud AI. As you read this book, the author first introduces the basics of cloud computing and then discusses the detailed topics of AI integration and automation, so that readers have a systematic understanding of current cloud systems.

In every chapter, I have endeavoured to explain technical concepts as simply as possible and repeatedly stressed on use cases in industries including healthcare, financial, manufacturing, retail, and smart city. This book provides information that will be valuable no matter if you are working in cloud architecture, data science or business management to achieve scalable implementation of AI and remarkable automation.

It is from such backgrounds that this book has been informed by the need to provide more light given the fast-growing adoption of technologies. Here are my objectives: To share practical strategies, not only abstract concepts so that the readers can learn not only concepts of cloud-AI integration but also learn how to make it work for their organization.

Indeed, it is my earnest desire that this book shall be useful to you as you strive to gain mastery over cloud technologies and to gain a release from effort-headache through artificial intelligence and automation.

DEDICATION

To my Family, well-wishers and young learners who inspire and challenge me to contribute to the ever-evolving journey of learning and innovation.

ACKNOWLEDGEMENT

Working on the book titled "Cloud Technologies with a Focus on AI Integration and Automation" has been incredibly enriching, and there are few people to whom I owe my heartfelt thankfulness.

Heartfelt gratitude to my family and friends because of their support and patience with me throughout the development of this work. The belief that my work is worth it from them has been encouraging all this time.

To fellow professionals and from whom I have been able to learn a lot, your input in shaping this work is highly appreciated. I am particularly grateful for the information sharing that defined the content and the case samples included in this work.

A special credit goes to the pathfinders in the sphere cloud computing Artificial intelligence and automation whose inventions are still encouraging the field. You have made the way for such technologies which this book speaks about it making, to be realized possible.

Finally, to the readers—it is really your curiosity and desire and desire to discover new things in the intersection of cloud, AI, and automation that drives further development of this topic. I appreciate your decision about selecting this book for your further education.

With sincere appreciation,

[Vidyasagar Vangala]

CONTENTS

Preface 3

Dedication 5

Acknowledgement 6

1. Introduction to Cloud Technologies and AI Integration 13

  1.1 Chapter Overview 13

  1.2 Cloud Computing: Concepts and Architecture 13

    1.2.1 Cloud Computing Architecture 14

    1.2.2 Components of Cloud Computing Architecture 16

    1.2.3 Benefits of Cloud Computing Architecture 17

  1.3 The Evolution of Cloud Technologies 18

  1.4 Types of Cloud Deployments: Public, Private, Hybrid, and Multi-Cloud 21

    1.4.1 Cloud Deployment Models 21

  1.5 Security in the Cloud: Best Practices and Frameworks 30

    1.5.1 Best Practices for Cloud Security: 30

    1.5.2 Security Frameworks for Cloud Computing 32

  1.6 Scalability and Elasticity in Cloud Solutions 36

    1.6.1 Cloud Elasticity 36

    1.6.2 Cloud Scalability 37

1.7 Overview of Cloud Service Models: IaaS, PaaS, and SaaS   40

    1.7.1 IaaS   41

    1.7.2 PaaS   42

    1.7.3 SaaS   44

1.8 Emerging Technologies Shaping Cloud and AI   46

1.9 Chapter Summary   48

Multiple Choice Questions (MCQs)   48

2. AI Integration in Cloud Computing   52

2.1 Chapter Overview   52

2.2 AI as a Service (AIaaS): A New Frontier   52

2.3 Machine Learning and Deep Learning in
Cloud Platforms   59

2.4 Data Management for AI-Driven Cloud Solutions   65

2.5 Natural Language Processing in Cloud-Based
Applications   69

2.6 Cloud AI Tools and Frameworks   73

2.7 Chapter Summary   75

Multiple Choice Questions (MCQs)   75

3. Automation in the Cloud Era   79

3.1 Chapter Overview   79

3.2 Automation in Cloud Contexts   79

3.3 DevOps and Continuous
Integration/Continuous Deployment (CI/CD)   82

3.4 Infrastructure as Code (IaC): Simplifying Cloud
Management   92

3.5 AI-Powered Automation: Predictive Maintenance and
Self-Healing Systems 115

3.6 Chatbots and Virtual Assistants in Automated
Cloud Services 118

3.7 Chapter Summary 128

Multiple Choice Questions (MCQs) 128

**4. Cloud Security and Compliance in AI-Driven
Environments** **132**

4.1 Chapter Overview 132

4.2 Cloud Security Architecture and Threat Mitigation 133

4.2.1 Cloud Security and Shared Responsibility Model 136

4.2.2 Principles of Cloud Security Architecture 137

4.2.3 Cloud Security Architecture Threats 138

4.3 Data Privacy and Compliance Regulations 140

4.4 Secure AI Workflows in Cloud Environments 148

4.5 Identity and Access Management (IAM) Best Practices 151

4.6 Cloud-Native Security Tools and Frameworks 158

4.7 Chapter Summary 165

Multiple Choice Questions (MCQs) 165

**5. Scalability and Performance Optimization in
Cloud-AI Systems** **168**

5.1 Chapter Overview 168

5.2 Designing Scalable Cloud Architectures 169

5.2.1 Building Scalable Cloud Architecture 172

5.3 Elasticity and Load Balancing in AI Workflows 178

5.4 Performance Monitoring and Optimization Tools   185

5.5 Cost Management and Resource Allocation Strategies   198

5.6 Real-World Examples of Scalable Cloud-AI
Implementations   204

5.7 Chapter Summary   207

Multiple Choice Questions (MCQs)   *207*

**6. Industry Applications of Cloud-AI Integration and
Automation**   **211**

6.1 Chapter Overview   211

6.2 AI-Driven Cloud Solutions in Healthcare   211

6.3 Revolutionizing Financial Services with
Cloud-AI Automation   218

6.4 Smart Cities: IoT, AI, and the Cloud Convergence   222

6.5 E-commerce and Retail: Personalization Through
Cloud AI   225

6.6 Manufacturing: Cloud and AI in Industry 4.0   227

6.7 Education: Enhancing Learning Through
Cloud-AI Tools   229

6.8 Media and Entertainment: Cloud AI for Content
Creation and Delivery   232

6.9 Transportation and Logistics: Autonomous Systems and
Cloud Integration   234

6.10 Chapter Summary   236

Multiple Choice Questions (MCQs)   *237*

7. Future Trends and Ethical Considerations in Cloud-AI Technologies **240**

7.1 Chapter Overview 240

7.2 Building Resilient and Scalable Cloud-AI Infrastructures 240

7.3 AI in Decentralized Cloud Systems: Blockchain and Beyond 246

7.4 Green Cloud Computing: Sustainability Through AI and Automation 249

7.5 Ethical and Societal Implications of AI in the Cloud 255

7.6 Future Innovations Shaping Cloud and AI Integration 259

7.7 Chapter Summary 262

Multiple Choice Questions (MCQs) *263*

Bibliography *267*

About the Author *277*

Chapter 01

INTRODUCTION TO CLOUD TECHNOLOGIES AND AI INTEGRATION

1.1 Chapter Overview

This chapter provides a foundational understanding of cloud technologies and their convergence with AI integration, setting the stage for deeper exploration in subsequent chapters. It introduces key concepts such as cloud computing architectures, service models (IaaS, PaaS, SaaS), and deployment types, emphasizing how these frameworks enable scalable, secure, and flexible computing environments. The chapter also highlights the rapid evolution of cloud technologies and how emerging innovations, such as AI-driven automation and predictive analytics, are transforming industries. By examining the role of security, scalability, and elasticity in modern cloud solutions, this chapter lays the groundwork for understanding the strategic advantages of combining cloud technologies with AI capabilities.

1.2 Cloud Computing: Concepts and Architecture

One of the most demanding technologies available today, cloud computing is giving every organisation a new form by offering on-demand virtualised services and resources. All sizes of businesses, from startups to major corporations, use cloud computing services to store data and retrieve it at any time and from any location using just the internet. We shall discover more about cloud computing's core architecture in this essay (Jadeja & Modi, 2012)the internet has changed the computing world in a drastic way. It has traveled from the concept of parallel computing to distributed computing to grid computing and recently to cloud computing. Although the idea of cloud computing has been around for quite some time, it is an emerging field of computer science. Cloud computing can be defined as a computing environment where computing needs by one party can be outsourced to another party and when need be arise to use the computing power or resources like database or emails, they can access them via internet. Cloud computing is a recent trend in IT that moves computing and data away from desktop and portable PCs into large data centers. The main advantage of cloud computing is that customers do not have to pay for infrastructure, its installation, required man power to handle such infrastructure and maintenance. In this paper we will discuss what makes all this possible, what is the architectural design of cloud computing and its applications. © 2012 IEEE.","author":[{"dropping-particle":"","family":"Jadeja","given":"Yashpalsinh","non-dropping-particle":"","parse-names":false,"suffix":""},{"dropping-particle":"","family":"Modi","given":"Kirit","non-dropping-particle":"","parse-names":false,"suffix":""}],"container-title":"2012 International Conference on Computing, Electronics and Electrical Technologies, ICCEET 2012","id":"ITEM-1","issued":{"date-parts":[["2012"]]},"title":"Cloud computing - Concepts, architecture and challenges","type":"paper-conference"},"uris":["http://www.mendeley.com/documents/?uuid=399b368f-78cc-421b-8106-3c890568ffcd"]}],"mendeley":{"formattedCitation":"(Jadeja & Modi, 2012.

1.2.1 Cloud Computing Architecture

Cloud computing architecture combines both Service Orientated Architecture (SOA) and Event Driven Architecture (EDA). The components of cloud computing architecture include storage, infrastructure, administration, security, client infrastructure, application, and runtime cloud.(Odun-Ayo et al., 2018).

The cloud architecture is divided into 2 parts, i.e.

1. Frontend

2. Backend

An interior architectural view of cloud computing is depicted in the image below.

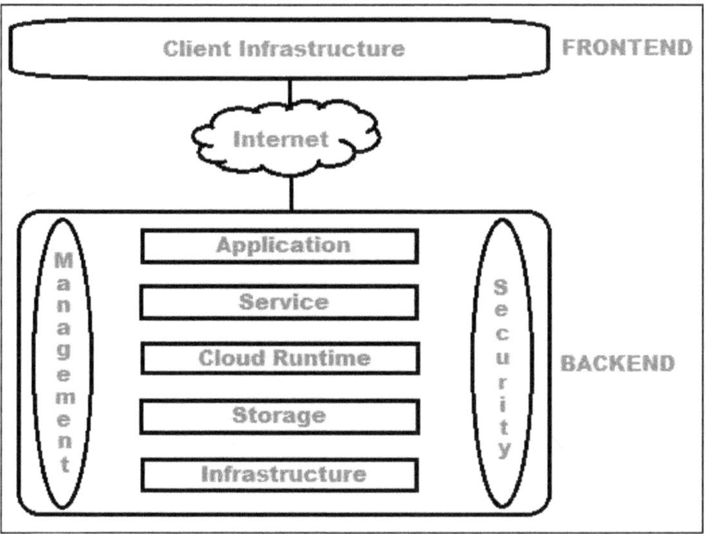

Source: - *(Geeksforgeeks, 2025)*

1. **Frontend:** The client side of the cloud computing system is referred to as the frontend of the cloud architecture. This indicates that it includes every application and user interface that the client uses to access cloud computing resources and services. For instance, accessing the cloud platform with a web browser.

2. **Backend:** The term "backend" describes the actual cloud that the service provider uses. In addition to managing the resources and offering security measures, it also contains them. Large storage,

virtual computers, virtual apps, traffic management systems, deployment methods, etc. are also included.

1.2.2 Components of Cloud Computing Architecture

The elements of the cloud computing architecture are as follows:

Client Infrastructure: A component of the frontend is the client infrastructure. It includes the user interfaces and applications needed to use the cloud platform. Stated differently, it offers a graphical user interface (GUI) for interacting with the cloud.

1. **Application:** An application is a type of backend component, which is a platform or piece of software that a client uses. This indicates that it offers the service in the backend in accordance with the needs of the customer.

2. **Service:** The three main categories of cloud-based services—SaaS, PaaS, and IaaS—are referred to as backend services. controls the kind of service the user accesses as well.

3. **Runtime Cloud:** The virtual machine's execution and runtime platform/environment are provided by the runtime cloud in the backend.

4. **Storage:** Backend storage offers scalable and adaptable storage services as well as data management.

5. **Infrastructure:** In the backend, "cloud infrastructure" refers to the hardware and software elements of the cloud, such as servers, storage, network equipment, virtualization software, etc.

6. **Management:** In backend management, backend components such as applications, services, runtime clouds, storage, infrastructure, and other security measures are managed.

7. **Security:** Implementing various security measures at the backend to safeguard end users' access to cloud resources, systems, files, and infrastructure is referred to as backend security.

8. **Internet:** An internet connection creates the interaction and communication between the frontend and backend by serving as a medium or bridge.

9. **Database:** Databases in the backend, such SQL and NOSQL databases, are used to store structured data. Google Cloud SQL, Microsoft Azure SQL, and Amazon RDS are a few examples of database services.

10. **Networking:** Backend networking services that offer networking infrastructure for cloud applications, including virtual private networks, load balancing, and DNS.

11. **Analytics:** Analytics in backend services that offer cloud data analytics features including machine learning, business intelligence, and warehousing.

A structural framework for developing, deploying, and overseeing cloud-based applications is offered by cloud computing architecture. The architecture of cloud computing offers advantages including cost-effectiveness, scalability, and adaptability. It also resolves issues with performance, dependability, and security.

1.2.3 Benefits of Cloud Computing Architecture

- Makes overall cloud computing system simpler.

- Improves data processing requirements.

- Helps in providing high security.

- Makes it more modularized.

- Results in better disaster recovery.

- Gives good user accessibility.

- Reduces IT operating costs.

- Provides high level reliability.

- Scalability.

1.3 The Evolution of Cloud Technologies

Numerous services that are kept on the Internet or in the cloud may be accessed by users thanks to cloud computing. Computer resources, data storage, servers, applications, development tools, and networking protocols are all included in cloud computing services. IT firms and businesses utilize it the most frequently(Borra, 2024).

Distributed computing gave rise to the contemporary technology known as cloud computing, which was initially referred to as "cloud computing" in the 1950s to represent internet-related services. Among the cloud services offered are those from Microsoft, Google, and Amazon. Many services that are kept on the Internet or in the cloud may be accessed by users thanks to cloud computing. Data storage, servers, applications, development tools, networking protocols, and computer resources are all included in cloud computing services(Kollolu, 2021).

The image below represents the evolution of Cloud Computing

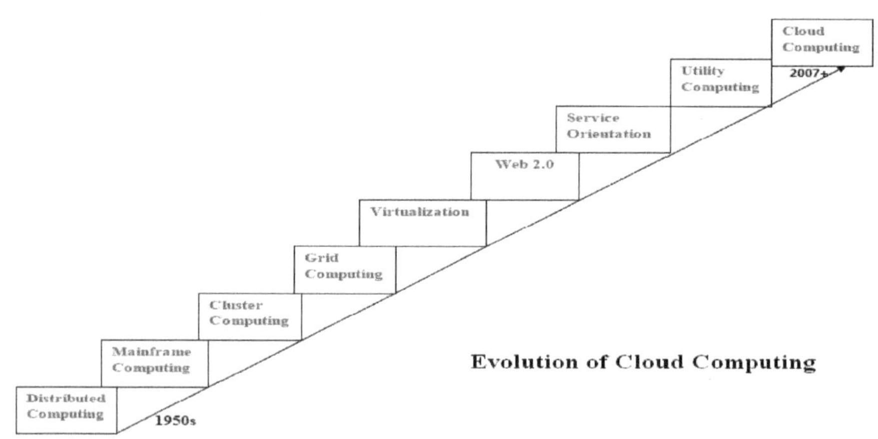

Source: - *(Geeksforgeeks, 2024)*

- **Distributed Systems:** A distributed system is made up of several separate systems that are all presented to consumers as one cohesive system. Distributed systems are designed to share resources and make effective and efficient use of them. Scalability, concurrency, continuous availability, heterogeneity, and failure-independent nature are some of the attributes of distributed systems. However, the primary issue with this system was that it required all of the systems to be present at the same place. In order to address this issue, distributed computing gave rise to three more computing paradigms: mainframe, cluster, and grid computing.

- **Mainframe Computing:** Mainframes are incredibly dependable and powerful computer devices that were originally introduced in 1951. These are in charge of managing big input-output activities and other large data. These are still utilised nowadays for large-scale processing jobs like internet transactions, etc. These systems have a high failure tolerance and almost little downtime. These improved the system's processing power after distributed computing. However, these were quite costly. Cluster computing emerged as a substitute for mainframe technology in order to lower this expense.

- **Cluster Computing:** Cluster computing emerged as a mainframe computer substitute in the 1980s. A high-bandwidth network linked every machine in the cluster to every other machine. Compared to those mainframe systems, they were far less expensive. These could perform complex calculations just as well. Additionally, if more nodes were needed, they could be added to the cluster with ease. Therefore, the cost issue was partially resolved, but the issue with regional limitations remained. The idea of grid computing was presented as a solution.

- **Grid Computing:** The idea of grid computing was first presented in the 1990s. It indicates that many systems were installed in completely disparate geographic places and were all connected via the Internet. Because these systems were owned by various organisations, the grid's nodes were diverse. Even while it resolved certain issues, when the distance between the nodes grew, other issues surfaced. The primary issue that was faced was the limited availability of high-speed connectivity, along with other challenges related to the network. As a result, cloud computing is frequently called the "Successor of Grid Computing".

- **Virtualization:** Virtualization first appeared around 40 years ago. It describes the procedure of building a virtual layer on top of the hardware that enables the user to run several instances on the hardware at once. It is a crucial piece of cloud computing technology. It serves as the foundation for popular cloud computing services like VMware vCloud, Amazon EC2, and others. One of the most popular forms of virtualization is still hardware virtualization.

- **Web 2.0:** The interface that cloud computing firms use to communicate with their customers is called Web 2.0. Our dynamic and interactive web pages are a result of Web 2.0. Additionally, it makes web pages more flexible. Web 2.0 is exemplified by popular sites like Facebook, Twitter, and Google Maps. It goes without saying that this technology alone makes social media feasible. In 2004, it became quite popular.

- **Service Orientation:** A service orientation serves as a cloud computing reference model. It facilitates applications that are affordable, adaptable, and dynamic. This computer model introduced two significant ideas. Software as a Service (SaaS) and Quality of Service (QoS), which include the SLA (Service Level Agreement), were among them.

- **Utility Computing:** A computing paradigm known as utility computing outlines methods for providing compute services as well as other important services like storage, infrastructure, etc. that are provided on a pay-per-use basis.

- **Cloud Computing:** Cloud computing is the practice of storing and gaining access to data and applications on distant servers hosted via the internet rather than on a local server or the computer's hard drive. Cloud computing, sometimes called Internet-based computing, is a technology that allows users to access resources as a service via the

Internet. Files, pictures, papers, and other storable materials can all be considered data.

Advantages of Cloud Computing

- Cost Saving
- Data Redundancy and Replication
- Ransomware/Malware Protection
- Flexibility
- Reliability
- High Accessibility
- Scalable

Disadvantages of Cloud Computing

- Internet Dependency
- Issues in Security and Privacy
- Data Breaches
- Limitations on Control

In conclusion, cloud computing, which provides exceptional flexibility, scalability, and efficiency, has grown to be a significant part of contemporary IT infrastructure. In addition to overcoming the limitations of earlier computer models, it has paved the way for new services and applications that will continue to advance economic development and technical advancement.

1.4 Types of Cloud Deployments: Public, Private, Hybrid, and Multi-Cloud

1.4.1 Cloud Deployment Models

A shared pool of computer resources, such as servers, storage, and apps, are made available to consumers through cloud computing. Flexible scalability is possible by requesting more resources as needed. Cloud platforms enable rapid deployment of resources, simplifying the process of getting systems operational. Unused resources can be released when no longer required, promoting cost efficiency through a pay-as-you-go model. The maintenance and management of these resources are handled entirely by the cloud service provider (Falatah & Batarfi, 2014).

What is a Cloud Deployment Model?

Depending on how much data you wish to keep and who has access to the infrastructure, the Cloud Deployment Model operates as a virtual computing environment with a deployment architecture that changes.

Types of Cloud Computing Deployment Models

Based on the nature and purpose of the cloud, as well as ownership, size, and access, the cloud deployment model determines the precise kind of cloud environment. A cloud deployment paradigm determines the location of the servers you're using and who has management over them. What you can alter, how your cloud architecture will seem, and if you will receive services or have to build everything yourself are all covered. Cloud deployment types also dictate how your users and the infrastructure interact. The following describes several kinds of cloud computing deployment models.

1. **Public Cloud**

    The public cloud enables anybody to access services and systems. Due to its open nature, the public cloud could be less secure. Cloud infrastructure services are made available to the general public or significant business organisations via the internet in a public cloud. The company that provides the cloud services, not the customer, owns the infrastructure under this cloud model. Accessing systems and services is made simple for users and clients by this kind of cloud hosting. An outstanding illustration of cloud hosting is this type of cloud computing, where service providers provide their services to a diverse clientele. Backup and retrieval services for storage are provided under this arrangement either for free, as a subscription, or on a per-user basis. Consider Google App Engine, for instance.

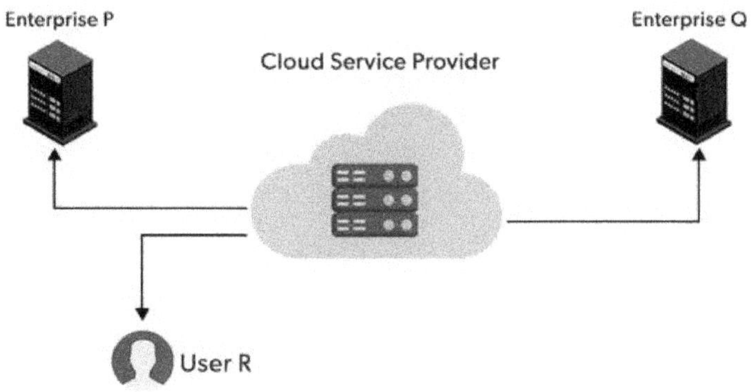

Figure 1.1: Public Cloud

Source: - *(Geeksforgeeks, 2023a)*

Advantages of the Public Cloud Model

- **Minimal Investment:** Because it is a pay-per-use service, there isn't a large upfront cost, which makes it ideal for businesses that need resources right now.

- **No setup cost:** There is no need to install any hardware because cloud service providers fully subsidise the entire infrastructure.

- **Infrastructure Management is not required:** It is not necessary to manage infrastructure in order to use the public cloud.

- **No maintenance:** The service provider performs the maintenance, not the users.

- **Dynamic Scalability:** Resources are available on demand to meet your business's needs.

Disadvantages of the Public Cloud Model

- **Less secure:** Public clouds are less secure since there is no assurance of high-level security because the resources are shared.

- **Low customization:** Since a large number of people may access it, it cannot be tailored to meet individual needs.

2. **Private Cloud**

The public cloud deployment paradigm and the private cloud deployment approach are diametrically opposed. For one user (client), it's a one-on-one setting. Sharing your gear with others is not necessary. How you manage all of the hardware is where private and public clouds differ from one another. The capacity to access systems and services within a certain organisation or border is referred to as the "internal cloud." The cloud platform is put into use in a secure cloud environment that is guarded by strong firewalls and managed by the IT department of a company. More control over cloud resources is possible with the private cloud.

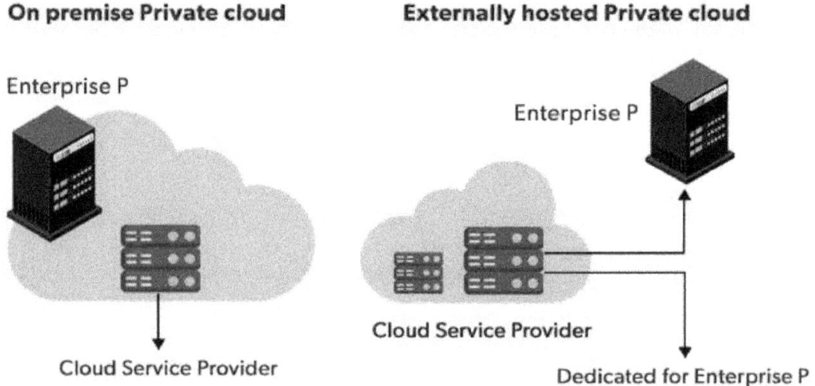

Figure 1.2: Private Cloud

Source: - *(Geeksforgeeks, 2023a)*

Advantages of the Private Cloud Model

- **Better Control:** The property belongs to you alone. You acquire total control over IT operations, rules, user behaviour, and service integration.

- **Data Security and Privacy:** It works well for keeping company data that only authorised employees may access. Resources can be divided out within the same infrastructure to increase security and accessibility.

- **Supports Legacy Systems:** This method is intended to be used with legacy systems that cannot connect to the public cloud.

- **Customization:** A private cloud deployment, as opposed to a public one, enables an organisation to customize its solution to match its unique requirements.

Disadvantages of the Private Cloud Model

- **Less scalable:** Since private clouds serve fewer clients, they are scalable within a specific range.

- **Costly:** Private clouds are more expensive since they offer individualized services.

3. **Hybrid Cloud**

Hybrid cloud computing offers the best of both worlds by utilising proprietary software to bridge the gap between the public and private spheres. With a hybrid solution, you may benefit from the cost advantages of the public cloud while hosting the app in a secure setting. Depending on their requirements, businesses might use a mix of two or more cloud deployment techniques to transfer data and apps between clouds.

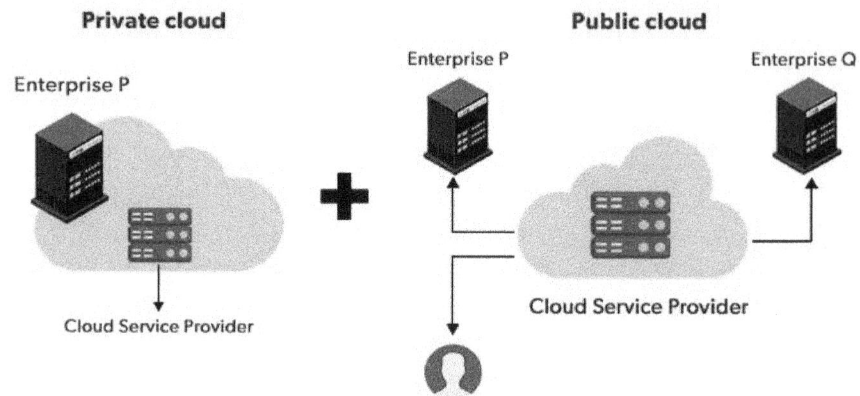

Figure 1.3: Hybrid Cloud

Source: - *(Geeksforgeeks, 2023a)*

Advantages of the Hybrid Cloud Model

- **Flexibility and control:** Greater flexibility allows businesses to create customized solutions that are tailored to their specific requirements.

- **Cost:** Scalability is a feature of public clouds, so you will only have to pay for more capacity if you need it.

- **Security:** The likelihood of data theft by attackers is significantly decreased when data is appropriately divided.

Disadvantages of the Hybrid Cloud Model

- **Difficult to manage:** Managing hybrid clouds, which include public and private cloud services, may be challenging. So, it's complicated.

- **Slow data transmission:** In the hybrid cloud, delay arises because data transmission occurs via the public cloud.

4. **Community Cloud**

It enables a collection of organisations to access systems and services. It is a distributed system designed to meet the unique requirements of a community, company, or industry by combining the services of several clouds. Organisations with similar issues or responsibilities may share the community's infrastructure. Usually, one or more community organisations work together to administer it, or a third party does.

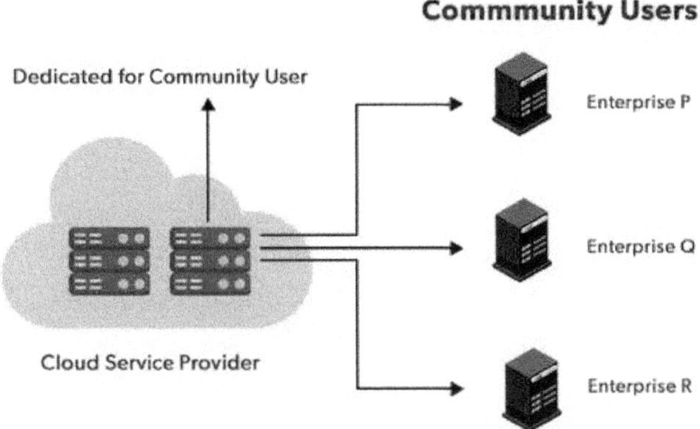

Figure 1.4: Community Cloud

Source: - *(Geeksforgeeks, 2023a)*

Advantages of the Community Cloud Model

- **Cost Effective:** Since many communities or organisations share the cloud, it is affordable.

- **Security:** Better security is offered by the community cloud.

- **Shared resources:** It enables you to share infrastructure, resources, and other things with other organisations.

- **Collaboration and data sharing:** It works well for data exchange and teamwork.

Disadvantages of the Community Cloud Model

- **Limited Scalability:** Since numerous organisations share resources based on their shared interests, community clouds are comparatively less scalable.

- **Rigid in customization:** Since data and resources are shared across several organisations based on their common interests, one organisation cannot make adjustments to meet its needs because doing so will affect other organisations.

5. Multi-Cloud

As the name suggests, we are discussing the use of several cloud providers simultaneously under this paradigm. It is comparable to the hybrid cloud deployment strategy, which blends resources from the public and private clouds. Multi-cloud makes use of several public clouds rather than combining private and public ones. Accidents can happen even if public cloud companies provide a variety of methods to increase the dependability of their services. It is extremely uncommon for two different clouds to experience an occurrence simultaneously. Therefore, multi-cloud implementation further enhances your services' high availability.

Figure 1.5: Multi-Cloud

Source: - *(Geeksforgeeks, 2023a)*

Advantages of the Multi-Cloud Model

- Selecting many cloud providers allows you to combine the finest aspects of each service to meet the needs of your workloads, apps, and company.

- **Reduced Latency:** Cloud locations and zones near your customers can be selected to lower latency and enhance user experience.

- **High availability of service:** It is extremely uncommon for two different clouds to experience an occurrence simultaneously. Thus, multi-cloud deployment enhances your services' high availability.

Disadvantages of the Multi-Cloud Model

- **Complex:** The system is complicated by the integration of several clouds, and bottlenecks might arise.

- **Security issue:** The intricate structure may contain vulnerabilities that a hacker may exploit, rendering the data vulnerable.

The different types of cloud deployment models include; Public, Private, Hybrid, Community and Multi-Cloud, all of which have their benefits and drawbacks depending on; control, security, scalability, and costs. Public clouds are cheaper, available as a service with access granted on demand but they don't have high security measures, on the other hand private clouds are more controlled and secure but expensive. This is an excellent blend between the two concepts since it is able to take advantages of both private and public clouds. Community cloud helps the organizations that share similar requirements to share the resource of the cloud while the multi-cloud strategy means that resources are spread across the different providers to enhance reliability and minimize latency. By familiarising with these deployment types, it is possible for any business to choose the right cloud model that meets the operational and security needs of an organisation.

1.5 Security in the Cloud: Best Practices and Frameworks

A key component of contemporary cloud computing is cloud security, which focusses on protecting infrastructure, data, and apps from cyberattacks, illegal access, and data breaches. Ensuring strong security procedures is essential to preserving data integrity, operational continuity, and regulatory compliance as more and more businesses move to cloud environments. Proactive security measures are crucial for safeguarding sensitive data and vital business activities in cloud settings because of the shared infrastructure, remote access, and dynamic nature of cloud services.

1.5.1 Best Practices for Cloud Security:

1. **Data Encryption:** Data encryption is essential for protecting private data kept on cloud servers. It entails transforming data into a safe format that requires the right decryption key to decode. Encrypting data while it's in transit (during data transmission) and at rest (when stored) makes sure that unauthorized people can't access it, even if they manage to get into the communication or storage channels. Because of their robust security features, modern encryption systems like AES-256 (Advanced Encryption Standard) are frequently employed. Furthermore, end-to-end encryption guarantees that data is safe at every stage of its existence, from creation to destruction.

2. **Identity and Access Management (IAM):** To regulate who has access to cloud resources and data, a robust Identity and Access Management (IAM) system must be put in place. By offering methods for controlling user identities, roles, and permissions, IAM frameworks make sure that only those with the proper authorization may access certain data or systems. The Principle of Least Privilege (PoLP) and Role-Based Access Control (RBAC) further reduce security threats by limiting user access to only the resources required for their job duties. A password and a one-time authentication code

are only two of the many types of verification that Multi-Factor Authentication (MFA) requires, adding an additional layer of protection.

3. **Regular Security Audits and Vulnerability Assessments:** Regular security audits assist businesses in locating and addressing cloud infrastructure problems. At regular evaluations, configuration errors, out-of-date software, and other security flaws that an attacker may exploit are scanned for. Evaluate the efficacy of current security measures by simulating real-world assaults through vulnerability assessments and penetration testing (ethical hacking). Real-time insights into security events and possible threats are obtained by continuous monitoring using Security Information and Event Management (SIEM) systems, allowing for proactive threat mitigation.

4. **Secure APIs and Endpoints:** Cloud services often rely on Application Programming Interfaces (APIs) to enable interaction between applications and cloud resources. Insecure APIs can become a major vulnerability if not properly secured. Implementing API gateways, strong authentication protocols (like OAuth2), and encryption of API communications ensures secure data exchange. Rate restriction, input validation, and the usage of Web Application Firewalls (WAFs) to identify and stop malicious activity directed at APIs are further components of proper API security.

5. **Backup and Disaster Recovery:** In the case of data loss, cyberattacks, or natural catastrophes, company continuity depends on a thorough backup and disaster recovery plan. To avoid single points of failure, cloud infrastructures should provide globally dispersed data storage and automatic, regularly scheduled backups. Scalable solutions are offered by Disaster Recovery as a Service (DRaaS) to reduce downtime and promptly restore data and services following an incident. Disaster recovery strategies should be tested often to guarantee readiness for unforeseen circumstances.

1.5.2 Security Frameworks for Cloud Computing

Security frameworks provide structured guidelines for implementing and maintaining secure cloud environments. These frameworks create uniform security procedures for cloud operations, enhance risk management, and guarantee adherence to industry standards.

NIST Cybersecurity Framework (CSF): The Cybersecurity Framework, created by the National Institute of Standards and Technology (NIST), offers a thorough set of recommendations for controlling cybersecurity threats in cloud systems. Identify, Protect, Detect, Respond, and Recover are its five main pillars. This framework helps organizations establish a proactive security posture by focusing on risk assessment, continuous monitoring, and incident response planning.

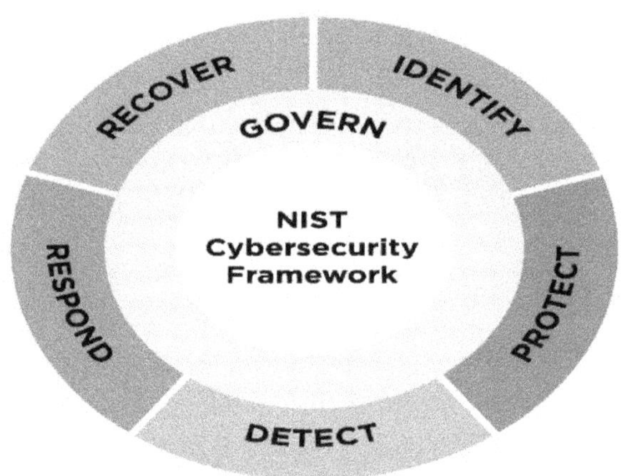

Figure 1.6: NIST Cybersecurity Framework (CSF)

Source: - *(Boutin, 2023)*

The "govern" role, which NIST has added to the five primary pillars of a successful cybersecurity program, highlights that cybersecurity is a significant source of corporate risk and that senior leadership should take it into account.

ISO/IEC 27001 – Information Security Management System (ISMS): An globally accepted standard for information security management is ISO/IEC 27001. It offers a thorough collection of rules, processes, and controls that offer a methodical approach to protecting sensitive data. Organisations may guarantee data availability, confidentiality, and integrity by putting ISO/IEC 27001 into practice. The standard also emphasizes risk assessment, continuous improvement, and regulatory compliance, making it ideal for securing cloud operations.

Figure 1.7: A globally accepted standard for information security management systems (ISMS) is ISO 27001. It gives businesses a framework for creating, putting into practice, maintaining, and continuously enhancing their information security management procedures.

Source: - *(Indiamart, 2023)*

CIS Controls (Centre for Internet Security Controls): The CIS Controls are a collection of best practices and prioritized activities intended to shield enterprises from the most frequent cybersecurity risks. These controls concentrate on things like monitoring privileged access, secure setups, and ongoing vulnerability management. Because CIS Controls provide doable actions to enhance security hygiene and stop typical attack routes, they are very useful in cloud contexts.

Figure 1.8: The 18 CIS controls

Source: - *(Impact, 2024)*

The 18 CIS Controls, a set of cybersecurity best practices created by the Centre for Internet Security, are shown in the above figure. A prominent lock sign serves as the focal point of the controls, highlighting the necessity of a multi-layered defines-in-depth approach. Asset inventory, data protection, account management, access control, vulnerability management, malware defences, and incident response are just a few examples of the crucial areas for security attention that each numbered control represents. These controls are further grouped into three implementation levels, with Level 1 being the most fundamental and Level 3 being the most advanced. The image provides a visual roadmap for organizations to improve their cybersecurity posture by implementing these controls, reducing their risk of cyberattacks and protecting their valuable assets (NIST, 2018).

Cloud Security Alliance (CSA) STAR Framework: A security architecture created especially for cloud service providers and customers is provided by the Cloud Security Alliance's (CSA) Security, Trust, Assurance, and Risk (STAR) initiative. The CSA STAR framework focuses on transparency, trust, and accountability in cloud operations. It provides guidelines for risk assessment, compliance management, and security automation, helping organizations build trust with their customers while maintaining a secure cloud environment.

Figure 1.9: The image is the logo for the Cloud Security Alliance's STAR (Security, Trust, and Assurance Registry) Enabled Solution.

Source: - *(CSA STAR Registry, 2022)*

Maintaining regulatory compliance, preserving sensitive data, and securing corporate operations all depend on strong security in cloud settings. Businesses may improve their cloud security posture by using best practices including data encryption, IAM, frequent audits, and disaster recovery planning. Additionally, implementing security frameworks like NIST CSF, ISO/IEC 27001, CIS Controls, and CSA STAR provides structured guidance for securing cloud infrastructures. Maintaining a safe and robust cloud environment will depend on being proactive with security frameworks and tactics as cloud technologies continue to advance.

1.6 Scalability and Elasticity in Cloud Solutions

1.6.1 Cloud Elasticity

The capacity of a cloud to dynamically expand or compress its infrastructure resources in response to abrupt changes in demand so that workload may be effectively managed is known as elasticity. The cost of infrastructure is reduced thanks to this flexibility. This isn't useful for every type of setting; it's best to focus on situations when resource needs abruptly change over a certain period of time. Using it in situations where a permanent resource infrastructure is needed to manage the high workload is not entirely feasible. For corporate or mission-based applications, where even a small change in display might result in significant financial loss, adaptability is essential. Therefore, in order to satisfy the presentation requirements, flexibility is required, and additional resources are allocated for such an application.

It functions so that as the number of clients increases, applications automatically allocate additional resources for figuring, stockpiling, and organizing, such as the central processor, memory, stockpiling, or transfer speed. Conversely, when the number of clients decreases, it automatically reduces those as needed. One well-known feature of cloud computing is its flexibility, which is linked to scale-out arrangements (level scaling), which consider assets that may be effectively added or removed as needed. Mostly, it has to do with public cloud resources, which are often emphasized in pay-per-use or pay-more-only when expenses occur solutions.

Flexibility is the ability to build or contract framework assets (such as process, capacity, or organisation) effectively on a case-by-case basis in order to adapt autonomously to changes in application responsibilities. It makes the most drastic utilisation of assets, resulting in reserve cash for foundation expenses generally. Dependent on the environment, the framework's assets—which include more than just equipment, programming, networks, QoS, and other configurations—are subject to change.

The adaptability is entirely dependent on the environment as it might occasionally turn into a drawback if the execution of specific applications is likely guaranteed. It is most frequently utilised in public cloud services that charge by the usage. whereby IT administrators are only prepared to pay for the time that they used the resources.

1.6.2 Cloud Scalability

In order to manage the increasing workload, cloud scalability is utilised, where efficient software or application operation also requires high speed. Scalability is frequently employed when a steady deployment of resources is necessary to manage the burden statically.

Example: Imagine that you are the owner of a business with a small database in the beginning, but as your company grows over time, so does the size of your database. In this scenario, all you need to do is ask your cloud service provider to increase the database's capacity so that it can handle a high workload.

It is completely apart from everything you have read about Cloud Elasticity thus far. The organization's dynamic demands are met by elasticity, whilst its static needs are met by scalability. One such feature that the cloud offers is scalability, for which users must pay on a per-use basis. Therefore, in situations when the burden is considerable and grows steadily, scalability is helpful.

Types of Scalability:

a. Vertical Scalability (Scale-up):

This kind of scalability involves increasing the working environment's current resources' power in an upward way.

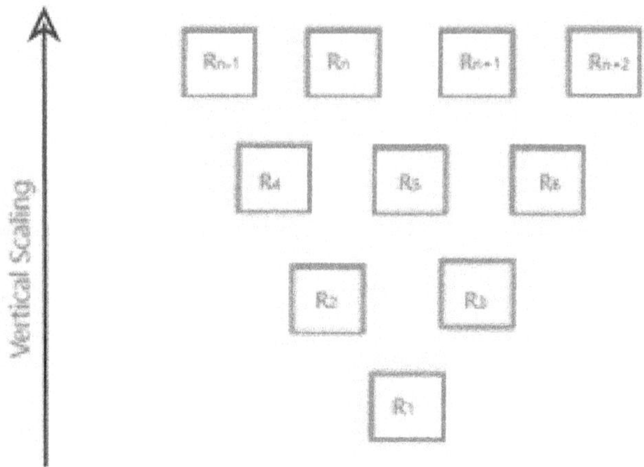

Figure 1.10: Vertical Scalability (Scale-up)

Source: - *(Geeksforgeeks, 2023b)*

b. Horizontal Scalability:

The resources are added in a horizontal row when scaling in this manner.

Figure 1.11: Horizontal Scalability

Source: - *(Geeksforgeeks, 2023b)*

c. Diagonal Scalability

The resources are added both horizontally and vertically, combining horizontal and vertical scalability.

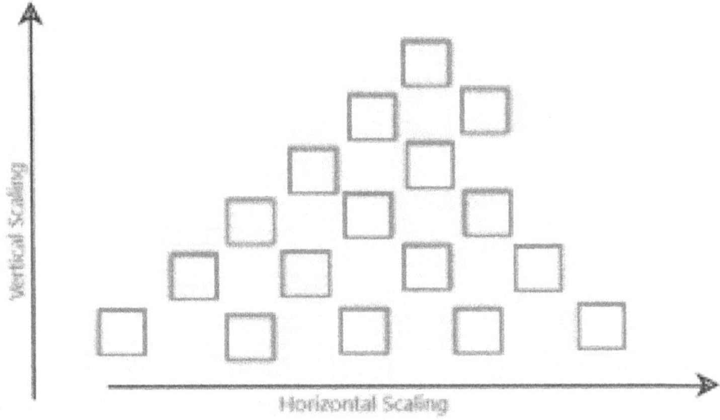

Figure 1.12: Diagonal Scalability

Source: - *(Geeksforgeeks, 2023b)*

The Distinction Between Scalability and Elasticity in the Cloud:

	Cloud Elasticity	**Cloud Scalability**
1	Elasticity is only utilised for a brief time to accommodate abrupt fluctuations in workload.	Scalability is employed to accommodate the workload's steady rise.
2	Elasticity is utilised to accommodate dynamic changes, such as an increase or reduction in the resources required.	Scalability is constantly utilised to deal with an organization's growing workload.
3	Small businesses that see temporary surges in demand and workload frequently deploy elasticity.	In order to operate efficiently, large corporations with a steadily expanding consumer base utilise scalability.

	Cloud Elasticity	Cloud Scalability
4	It is a short-term strategy that was implemented just to address unforeseen spikes in demand or seasonal needs.	Scalability is a long-term strategy that was implemented only to handle an anticipated rise in demand.

Source: - *(Geeksforgeeks, 2023b)*

1.7 Overview of Cloud Service Models: IaaS, PaaS, and SaaS

Cloud service models, which enable different degrees of control, administration, and flexibility depending on business needs, specify how cloud services are provided to customers. Each of the three main models—Software as a Service (SaaS), Platform as a Service (PaaS), and Infrastructure as a Service (IaaS)—offers unique advantages to businesses implementing cloud computing. IaaS gives companies control over infrastructure without requiring them to handle real hardware by providing virtualised computing resources including servers, storage, and networking. PaaS gives developers a platform to create, test, and launch apps while removing the complexity of the infrastructure and providing development and teamwork tools. SaaS is perfect for end customers looking for accessibility and convenience since it provides fully working software programs over the internet without the need for local installation or administration. These models empower organizations to choose the right balance between control, scalability, and simplicity based on their operational needs. IaaS, PaaS and SaaS are the three most popular types of cloud service offerings. They are sometimes referred to as cloud service models or cloud computing service models.

Infrastructure as a Service, or IaaS, is the backend IT infrastructure for executing workloads and applications in the cloud. It provides on-demand access to real and virtual servers, storage, and networking housed in the cloud.

- Platform as a Service, or PaaS, provides on-demand access to a comprehensive, usable, cloud-hosted platform for creating, executing, fixing, and overseeing applications.

- On-demand access to ready-to-use, cloud-hosted application software is known as software as a service, or SaaS.

1.7.1 IaaS

Customers can create, configure, and use cloud-hosted computing infrastructure (IaaS) on-demand, including servers, storage, and networking resources, in a manner similar to that of on-premises hardware(Microsoft, 2021).

The distinction is that the cloud service provider uses its own data centers to host, operate, and administer the computer resources and infrastructure. IaaS clients pay for the hardware's utilisation on a pay-as-you-go or subscription basis, and they do so over an internet connection. IaaS clients often have the option of bare metal servers on dedicated (unshared) physical hardware or virtual machines (VMs) hosted on shared physical hardware (virtualization is managed by the cloud service provider). Clients can use a graphical dashboard or application programming interfaces (APIs) to deploy, configure, and manage the servers and infrastructure resources. The initial 'as a service' offering is IaaS: IaaS was initially offered by all of the main cloud service providers, including Amazon Web Services, Google Cloud, IBM Cloud, and Microsoft Azure.

Benefits of IaaS

Customers have more flexibility with IaaS than with traditional IT as it allows them to scale computer resources up or down in response to traffic peaks or troughs. The upfront costs and overhead associated with buying and operating an on-premises data centre are avoided by clients using IaaS. It also removes the need to constantly balance the waste of buying extra on-premises capacity to handle spikes against the potential for poor performance or outages due to insufficient capacity for unforeseen traffic surges or bursts.

Other benefits of IaaS include:

- **Higher availability:** A business may quickly set up redundant servers using IaaS, even locating them in other locations to guarantee uptime in the event of a local power loss or natural disaster.

- **Lower latency, improved performance:** Apps and services can be located closer to consumers to reduce latency and maximize performance because IaaS providers usually run data centers across various locations.

- **Improved responsiveness:** In only a few minutes, customers may supply resources, test new concepts, and swiftly distribute them to other users.

IaaS use cases

Common uses of IaaS include:

- **Disaster recovery:** Rather than installing redundant servers in many places, IaaS may integrate its disaster recovery solution into the cloud provider's already-existing globally scattered infrastructure.

- **Ecommerce:** Online shops who see regular surges in traffic will find IaaS to be a great choice. In today's 24-hour retail sector, the capacity to expand during times of strong demand and superior security are crucial.

- **Internet of Things (IoT), event processing: artificial intelligence (AI):** Setting up and scaling up data storage and processing resources for these and other applications that handle massive amounts of data is made simpler by Infrastructure as a Service (IaaS).

1.7.2 PaaS

PaaS offers a cloud-based platform for creating, executing, and maintaining applications. All of the platform's hardware and software, including servers (for development, testing, and deployment), operating system (OS) software, storage, networking, databases, middleware, runtimes, frameworks, and development tools, are managed and maintained by the cloud services provider. Additional services include security, backups, and operating system and software upgrades (Pastore, 2013).

Developers or DevOps teams may work together on all aspects of the application lifecycle, including coding, integration, testing, delivery, deployment, and feedback, using a graphical user interface (GUI) to access the PaaS. AWS Elastic Beanstalk, Google App Engine, Microsoft Windows Azure, and Red Hat OpenShift on IBM Cloud are a few instances of PaaS technologies.

Benefits of PaaS

PaaS's main advantage is that it enables users to create, test, launch, update, and grow applications more rapidly and affordably than they could if they had to set up and maintain their own on-premises platform. Additional advantages consist of:

- **Faster time to market:** Instead of weeks or months, PaaS allows development teams to set up development, testing, and production environments in a matter of minutes.

- **Low- to no-risk testing and adoption of new technologies:** Access to a variety of the newest resources, both up and down the application stack, is usually provided by PaaS platforms. Without having to spend heavily in them or the equipment needed to operate them, this enables businesses to test new operating systems, languages, and other tools.

- **Simplified collaboration:** PaaS is a cloud-based service that offers a shared software development environment, enabling development and operations teams to use all necessary tools from any location with an Internet connection.

PaaS use cases

PaaS may support a number of IT and development projects, including:

- **API development and management:** PaaS facilitates the development, operation, management, and security of APIs for data and feature exchange across apps by providing teams with pre-built frameworks.

- **Internet of Things (IoT):** PaaS facilitates the development of IoT applications and the real-time processing of data from IoT devices by supporting a variety of programming languages (including Java, Python, Swift, and others), as well as tools and application environments.

- **Agile development and DevOps:** PaaS solutions usually have built-in automation to enable continuous integration and continuous delivery (CI/CD) and fulfil every need of a DevOps toolchain.

Cloud-native development and hybrid cloud strategy: Platform as a Service (PaaS) solutions facilitate cloud-native development technologies, such as microservices, containers, Kubernetes, and serverless computing, which allow developers to create once, deploy, and manage uniformly across on-premises, private, and public cloud environments.

1.7.3 SaaS

SaaS, also known as cloud application services, is application software that is hosted in the cloud and is ready for use. To utilize a full program from within a web browser, desktop client, or mobile app, users must pay a monthly or yearly price. The SaaS provider hosts and manages the application as well as all of the infrastructure needed to deliver it, including servers, storage, networking, middleware, application software, and data storage.(Kavitha & Damodharan, 2020).

All software updates and upgrades are handled by the vendor, typically without the consumers' knowledge. As part of a service level agreement (SLA), the vendor usually guarantees a certain degree of performance, security, and availability. On demand, customers may add extra users and data storage for a fee.

SaaS is virtually a given for everyone using a mobile phone these days. SaaS apps that individuals use on a daily basis in their personal life include social networking, email, and cloud file storage software like Dropbox or Box.

The following are well-known business or enterprise SaaS solutions: Canva (graphics), Slack (collaboration and messaging), Trello (workflow management), Salesforce (customer relationship management software), and HubSpot (marketing software). There are currently several desktop apps (like Adobe Creative Suite) that are offered as software as a service (SaaS), like Adobe Creative Cloud.

Benefits of SaaS

The primary advantage of SaaS is that it transfers all application and infrastructure administration to the SaaS provider. The user only needs to register, pay the cost, and begin using the software. Everything else is handled by the vendor, including patching and upgrading the server, managing user access and security, storing and managing data, and more.

Other benefits of SaaS include:

- **Minimal risk:** A free trial period or inexpensive monthly costs are offered by many SaaS applications, allowing users to test the software and determine whether it meets their needs with little to no risk to their finances.

- **Anytime/anywhere productivity:** On any device with a browser and an internet connection, users may interact with SaaS apps.

- **Easy scalability:** Customers can purchase more data storage for a little fee, but adding users is as easy as registering and paying for additional seats.

Certain SaaS providers even allow for product customisation by offering a complementary PaaS solution. Heroku is a well-known example of a PaaS solution for Salesforce.

SaaS use cases

There are too many particular use cases to list them all, however SaaS is now accessible for almost any personal or employee productivity application (several are mentioned above). Most of the time, if a person or organisation can locate a SaaS option with the necessary capabilities, it will provide a far more straightforward, scalable, and economical substitute for on-premises software.

To sum up, the IaaS, PaaS and SaaS models provide organizations with both more flexible, scalable and less costly offerings while providing organizations with different levels of control and operation tasks. IaaS offers the physical underpinning of the infrastructure required by the organization, when an organization wants to manage their virtual resources directly, then PaaS offers a platform for developing applications where the development environment, testing, and deployment of applications are also managed fully. SaaS, in contrast, provides fully functional applications with little need for administration, SaaS is suitable for end-users and organizations that want simplicity. Combined, these models enable great strategies to be built that allow organizations to use cloud technologies more effectively, boosting productivity, developing and implementing innovation faster, and enabling digital transformation in today's highly competitive business environment.

1.8 Emerging Technologies Shaping Cloud and AI

New technologies are already changing the paradigm of the cloud and AI, pushing forward improvements in innovation, adaptability, and green performance across business verticals. Edge computing has emerged as a critical innovation that can support holistic data processing nearer to the source, and thereby minimizing response time for providing real-time artificial intelligence for intelligent transportation systems, integrated urban environments, and industrial Internet of Things. Besides, this shift reduces the amount of data sent to the central server hence boosting the speed needed and energy consumption.

Self-learning is also being revolutionised by quantum computing, as it provides previously unseen computational capabilities in speeding up some activities like deep learning, as well as pat-tern recognition and optimisation challenges which to date are restricted by recognised ideas. Quantum computers will inevitably find practical applications in solving problems in climatology, genetics and finance, as these technologies advance.

On the same note, serverless computing and containerization have disrupted cloud structures where developers can run applications with little or no control of the underlying servers. Serverless architectures are self-servicing and increase or decrease proportions depending on the volume of work, thus saving energy and costs. At the same time, containerization with Kubernetes and Docker helps improve application mobility and use of resources to address AI deployment challenges in hybrid and multi-cloud landscapes.

In addition, other cognitive automation tools like AIOps (Artificial Intelligence for IT Operations) is aiding in Cloud management by providing a way for monitoring systems, identifying abnormal operations, and adjusting them where necessary. These tools use machine learning methods for the prognosis of system failures, the efficient usage of resources and the provision of high availability of services thus improving both reliability and sustainability.

Another trend is energy-effective design and the use of renewable power sources as components of data centers that are used to support AI computations. Other advances such as liquid cooling systems and AI managed resources also play their part in green data processing.

Altogether, these terrific emerging technologies are changing the combined Cloud Computing & AI interface and enabling industries to transform toward more sustainable, efficient, and Intelligent Smart Digital World with the power of faster decision making, low operational expense, and an enhanced environmental impact.

1.9 Chapter Summary

This chapter is a brief introduction of the research topic, which was centred on the review of sustainable practices for data centre. It defined the new idea of renewable energy usage, energy efficient hardware, advanced cooling system, carbon footprint assessment and proper e-waste disposal. The chapter underlined the problem of the increasing energy consumption connected with the training of AI models and the necessity of the effective and environmentally friendly solutions. Furthermore, it described the research questions and objectives that would inform investigation into the ways that these sustainable practices can help to lessen carbon emissions and energy use. This summary is followed by subsequent chapters, which are devoted to the methodologies, analysis, and findings of the sustainability concern in data centers.

Multiple Choice Questions (MCQs)

1. **Which of the following best describes cloud computing?**

    a. Storing data on a personal computer

    b. Accessing and managing data and services over the internet

    c. Using software without an internet connection

    d. Running applications only on local servers

2. **Which cloud deployment model offers a mix of on-premises and third-party services?**

   a. Public Cloud

   b. Private Cloud

   c. Hybrid Cloud

   d. Multi-Cloud

3. **What is the primary advantage of using a private cloud?**

   a. Cost efficiency

   b. Greater control and security

   c. Global accessibility

   d. Ease of third-party collaboration

4. **Which of the following is an example of an Infrastructure as a Service (IaaS) solution?**

   a. Google Workspace

   b. Amazon EC2

   c. Microsoft Azure SQL Database

   d. Dropbox

5. **Elasticity in cloud computing refers to:**

   a. The ability to handle an increase in workload by dynamically allocating resources

   b. The ability to store unlimited data

   c. The ability to restrict user access

   d. The ability to reduce server performance

6. **Which of these bests describes a multi-cloud strategy?**

   a. Using multiple cloud services from different providers for various workloads

   b. Using a single cloud provider for all services

   c. Using only on-premises infrastructure

   d. Using a hybrid cloud with both public and private clouds

7. **Which of the following security practices is essential for cloud computing?**

   a. Ignoring encryption for speed optimization

   b. Implementing data encryption and access control

   c. Avoiding multi-factor authentication

   d. Sharing access credentials for easy collaboration

8. **Which of these technologies is considered an emerging trend in cloud and AI integration?**

   a. Serverless Computing

   b. Dial-up Internet

   c. Physical Data Centers

   d. Manual Data Processing

9. **What is the key feature of Platform as a Service (PaaS)?**

   a. Provides complete software solutions without the need for infrastructure management

   b. Offers a platform for developers to build applications without managing underlying hardware

   c. Provides raw computing power and storage capacity only

   d. Focuses on physical server deployment

10. The concept of cloud scalability refers to:

a. The ability to reduce computing resources

b. The ability to manage costs effectively

c. The ability to expand resources based on demand

d. The ability to limit access to cloud services

Answer

1	2	3	4	5	6	7	8	9	10
b	c	b	b	a	a	b	a	b	c

AI INTEGRATION IN CLOUD COMPUTING

2.1 Chapter Overview

This chapter discusses AI in the context of cloud computing while making conspicuous how AI technologies are revolutionizing cloud services and processes. It emphasizes the contribution of AI in strengthening automation, data handling, and decision-making prospects of cloud architectures for better productivity and capacity. The chapter also analyses the ways in which cloud platforms underpin the computational resources and storage essential for AI model training and deployment and thus make the technologies more readily available to organisations. In addition, it discusses the positive outcomes of this integration, including efficient resource utilization and enhanced data centre performance all-important and simultaneously discusses the problems of power consumption and sustainability. The series of explanation of how the AI-led inventions can be harmonised with an environmentally sensitive approach to cloud computing is introduced in this chapter.

2.2 AI as a Service (AIaaS): A New Frontier

AIaaS stands for Artificial Intelligence as a Service and it is another innovation in the idea of making the new technological advancement, AI, available to anyone through cloud service. AI as a Service enables companies to deploy machine learning, NLP, and computer vision applications that were previously impossible, or costly to implement, due to the high investments required in AI infrastructure. This model lowers the technical requirements needed for AI implementation and allows organizations of any scale to use data analytics and automations. In addition, the use of AIaaS means that organizations can scale up and down their AI resources more easily. With its further development, new possibilities for creating services and applications are appearing while issues of data protection, cybersecurity, and environmental management of resources in clouds are being provoked (Lins et al., 2021).

AI as a service A cloud-based service called artificial intelligence as a service (AIaaS) gives businesses access to artificial intelligence (AI) via a third-party product. This enables businesses to test AI for a variety of uses with less risk and without a significant upfront cost.

Many organizations find the upfront time and expense required to start a custom AI platform cost-prohibitive. An AIaaS platform, however, enables them to work with AI without these associated investments. AIaaS provides out-of-the-box platforms that are easy to set up, making it easier to test various public cloud platforms, services and machine learning (ML) algorithms.

How does AI work?

AI includes a wide range of technologies, such as natural language processing (NLP), computer vision, robotics, ML models, and cognitive computing.

The main AI tool, machine learning algorithms, are a set of rules or techniques that a computer often uses to calculate or resolve an issue. In order to solve issues or make decisions, computers often analyze large amounts of data or generate statistical projections and generalizations.

Artificial intelligence (AI) systems analyze vast volumes of training data, look for connections and patterns, and use these patterns to forecast future conditions. The training process typically requires massive amounts of data. The less data the model is trained on, the more likely it is to be biased.

Deep learning algorithms that employ deep neural networks and machine learning algorithms like regression and classification are the two main categories into which AI algorithms are commonly separated.

AIaaS provides access to already trained and existing AI models through the cloud. This means users can quickly scale resources up or down as needed. AIaaS platforms also commonly support different data sources, meaning an organization can integrate the service with its data environment. The AIaaS provider is responsible for managing model updates, infrastructure and security.

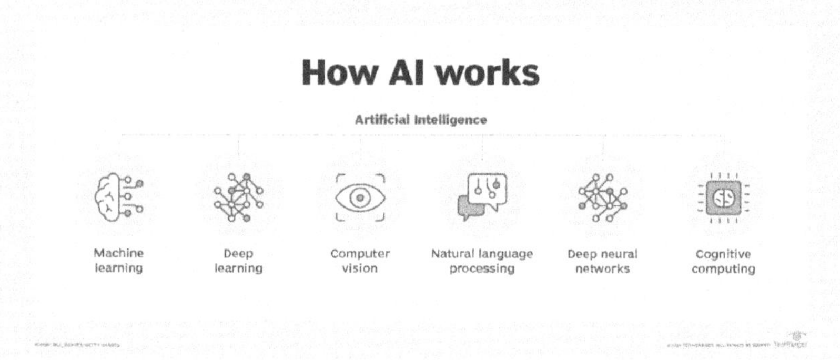

Figure 2.1: The figure depicts the different types of Artificial Intelligence (AI).

Source: - *(Gillis, 2024)*

The benefits of using AIaaS platforms

The AIaaS delivery model allows organisations to implement AI at a reasonable cost without having to create or manage a single AI project. AIaaS systems allow businesses to create Personalized, scalable, and user-friendly AI services.

The following are additional benefits of AIaaS systems:

- **Quick deployment**: AIaaS is one of the fastest ways to introduce AI to an organization. It's easy to install and set up. Because there are numerous AI use cases, it isn't always feasible for an organization to create and maintain an AI tool for each one. Customizable options are especially useful, as organizations can deploy AI services quickly and tweak them according to their business needs and constraints.

- **Low or no required coding skills**: AIaaS can be used even if a company lacks an in-house AI developer or programmer. All that's needed is a layer of no-code infrastructure in the enterprise, as generally, no coding or technical expertise is needed during the setup process.

- **Cost savings:** Saving money is the main factor influencing the expansion of AIaaS in the IT industry. AIaaS is cost-effective for businesses because they only pay for usage and AI functionality and don't need to make sizable upfront investments.

- **Price transparency:** In addition to reducing non-value-added labor, AIaaS also offers access to AI with a high level of transparency regarding service fees. Because most AIaaS pricing structures are based on consumption, businesses only pay for the AI technologies they use.

- **Scalability:** AIaaS is well suited for companies looking to scale. It's ideal for tasks that don't add significant value yet need some level of cognitive judgment. Because AIaaS uses industrial automation to complete simple tasks without requiring human intervention, team members have more time to focus on other tasks.

What are the challenges of AIaaS?

Although AIaaS offers many notable benefits, it also has the following challenges that organizations must consider:

- **Price:** AIaaS can be cost-prohibitive for many organizations, as the charges for usage and maintenance can be substantial over time. Organizations must evaluate their long-term objectives to ensure AIaaS aligns with their budget and overall goals.

- **Transparency:** Most AIaaS platforms provide users access to the services, but offer little to no transparency into their internal operations.

- **Security:** Data security is a major concern with AIaaS, as businesses must share data with outside vendors. However, data masking and other privacy-enhancing techniques are designed to safeguard an organization's data.

- **Data governance:** Businesses must tightly enforce limits on cloud data storage in highly regulated industries. For example, organizations in the banking and healthcare sectors might find AIaaS challenging to use because they could encounter restrictions on how data can be stored, shared and used in the AIaaS platform.

- **Vendor lock-in:** If one AIaaS provider isn't meeting a company's needs, it isn't easy to switch to another vendor. This is because various AI providers employ different response styles and vendor lock-in agreements. The transition might also be time-consuming for team members because they must learn the new program from scratch.

Types of AIaaS

Platforms for AI providers offers a variety of machine learning and AI approaches. Because they may test features and prices to determine what suits them, these variants can be tailored to an organization's AI needs. The specialised hardware required for some AI jobs, including graphics processing units for demanding workloads, can be provided by cloud AI service providers.

The following are different types of AIaaS:

- **Bots:** Chatbots and bots are used extensively in all sectors of the economy. They resemble actual human speech using natural language processing (NLP) and are typically employed in customer service to offer pertinent responses to the most common questions from clients. Companies save time and resources by responding around the clock and enabling employees to focus on more challenging tasks. A study conducted by AI provider Tidio found that 82% of customers would rather use a customer service chatbot than wait for human agents to respond to their inquiries.

- **Machine learning:** Companies use machine learning (ML) to analyze and spot patterns in their data, forecast outcomes, and gain knowledge along the way. This data processing technique is intended to run with little or no human intervention, empowering businesses to employ AIaaS without specialist technical skills. Pretrained models and models created for specific use cases are only two of the many alternatives available in machine learning.

- **Application programming interfaces:** An API is a software bridge that enables communication between two applications. An example of this is a third-party airline booking website -- such as Expedia, Kayak or Cheap air -- that uses information from several airline databases to display deals in one convenient location. APIs are also often used in machine vision, conversational AI, and natural language processing (NLP) applications including sentiment analysis and urgency detection.

- **Data labelling:** Data labelling is the process of annotating huge amounts of data to arrange it efficiently. It has numerous uses, such as guaranteeing data quality, categorizing data according to size and creating AI. The data is labelled using human-in-the-loop machine learning, which enables humans and machines to interact continuously and makes it easy for AI to evaluate the data in the future.

- **IoT:** Internet of Things (IoT) technologies, also known as AIoT, incorporate AI processes. The objective is to improve data management, analytics, and human-machine interactions by establishing more effective IoT operations. AI is integrated into hardware parts that are linked together via Internet of Things networks in AIoT devices. The AI evaluates the data that the devices generate and collect to offer insights that may be applied to increase productivity and efficiency.

Future of AIaaS

As the technology advances, businesses continue to find new applications for AI. Some common trends that are dictating the direction of AI and AIaaS include the following:

- **More humanlike conversations:** AI bots can handle more humanlike conversations. To further specialize chatbot use, unique personas can be created using data from user knowledge bases.

- **Increased customization:** AI services are also becoming more customizable, offering brands more options for personalizing responses based on previous user interactions and data collected.

- **Reduced need for large data silos:** Because vendors are training an AIaaS model, there's a reduced need for individual organizations to collect massive amounts of data. As AIaaS becomes more popular, fewer companies will need massive data sets for training purposes.

- **Better accessibility:** As AIaaS grows in popularity, it should gradually become more accessible to most businesses. No-code or low-code platforms should also increase the accessibility of AIaaS.

- **A focus on ethical AI:** There has been increasing controversy around how companies that are adopting AI technologies obtain their training data from users, AI biases, the environmental impact of AI or how AI might be implemented by companies as a cost-saving measure to help get rid of human jobs. This has sparked a growing conversation around ethical AI and how AI should be used.

Early adopters are drawn to AIaaS because it offers many benefits and is a rapidly expanding industry. AIaaS is expected to be as important as other as-a-service items, notwithstanding some obstacles to its advancement.

2.3 Machine Learning and Deep Learning in Cloud Platforms

Machine learning and deep learning are now critical parts of the contemporary cloud solution that allows organizations to leverage data-driven intelligence and task automation. Nowadays, cloud providers provide infrastructure for training and deploying ML and DL models, and ready environments, which significantly decreases the demand for own data centers and infrastructure. Some of these platforms offer sophisticated features which include; image recognition, voice and text analysis, predictive analysis and many others which makes the use of AI tools easier for businesses regardless of the industry they belong to. Thus, using cloud-based services in a case of ML and DL, organizations fasten up the pace of innovations, enhancement of decision-making, enhancement of operations' productivity, reserve resources, and manage costs efficiently (J. Rane et al., 2024).

What Is Machine Learning in the Cloud?

A kind of artificial intelligence called machine learning (ML) mimics human learning, enabling computers to become more predictive until they are able to carry out tasks without explicit programming. With the use of past training data, machine learning-driven software can forecast novel results. Large volumes of data, processing power, and infrastructure are needed to train an accurate machine learning model. Given the time and expense involved, most organisations find it challenging to train a machine learning model internally. The computation, storage, and services needed to train machine learning models are provided by a cloud ML platform.

Cloud computing enables developers to create machine learning algorithms more quickly while also making machine learning more affordable, accessible, and adaptable. An organisation may utilize pre-trained models for their applications (AI as a service) or select various cloud services to assist their ML training initiatives (GPU as a service), depending on the use case.

Machine Learning in the Cloud: Benefits and Limitations

Cloud-Based Machine Learning Advantages:

Many businesses may utilize open source frameworks like Scikit Learn, TensorFlow, or PyTorch to create machine learning models internally. Even if internal teams are competent in developing algorithms, they frequently struggle to scale models to real-world workloads and deploy them to production, which frequently calls for big computer clusters.

Deploying machine learning capabilities into corporate applications is hampered by a number of factors. The cost of manpower, development, and infrastructure is increased by the knowledge needed to create, train, and implement machine learning models. Additionally, specialised hardware equipment must be purchased and operated.

Cloud computing can help with many of these issues. Organisations may use machine learning skills to address business challenges without having to shoulder the technological load thanks to public clouds and AIaaS services.

The following is a summary of cloud computing's main advantages for machine learning workloads:

- On-demand pricing models enable ML efforts to be started without a significant upfront expenditure.

- The speed and performance of GPUs and FPGAs are available via the cloud without the need for a hardware investment.

- As projects enter production and the need for machine learning skills increases, the cloud enables companies to scale and experiment with these capabilities with ease.

- ML capabilities may be accessed through the cloud without the need for sophisticated data science or artificial intelligence expertise.

What Are the Limitations of Machine Learning in the Cloud?

Cloud-based machine learning has three main drawbacks:

- **Doesn't replace experts:** Even when cloud-managed, machine learning systems still need human oversight and improvement. The realistic limits of AI's capabilities without human supervision and involvement exist. Algorithms lack the knowledge necessary to comprehend a situation completely and to react appropriately to all potential inputs.

- **Data mobility:** Switching from cloud or service might be difficult when using machine learning models in the cloud. To achieve this, the data must be moved without compromising the model's functionality. Models for machine learning are frequently sensitive to even little modifications in the input data. A model could not function properly, for instance, if you need to alter the quantity or format of your data.

- **Security concerns:** The same issues that affect any cloud computing platform also affect cloud-based machine learning. Due to their frequent exposure to public networks, cloud-based machine learning systems are vulnerable to hacker intrusion, which might result in manipulated ML output or increased infrastructure expenses. Denial of service (DoS) attacks can potentially affect cloud-based machine learning models. Models implemented inside a corporate firewall are immune to many of these risks.

Types of Cloud-Based Machine Learning Services
Artificial Intelligence as a Service (AIaaS)

A delivery model known as Artificial Intelligence as a Service (AIaaS) allows vendors to offer artificial intelligence (AI) that lowers the risk and initial investment of its clients. Customers may test multiple machine learning (ML) algorithms and experiment with different cloud AI offerings by using the services that best fit their needs.

Every AIaaS provider provides a range of AI and machine learning services with varying capabilities and cost structures. For instance, several cloud AI providers provide specialised hardware, such as GPU as a Service (GPUaaS) for demanding workloads, for certain AI activities. AWS Sage Maker and other services offer a fully managed environment for developing and refining machine learning algorithms.

GPU as a Service (GPUaaS)

On-premises GPU infrastructure setup is no longer necessary thanks to GPU as a Service (GPUaaS) providers. GPU resources may be elastically provisioned on demand with these services. It increases scalability and flexibility, lowers the cost of in-house GPU infrastructure, and makes it possible for many to deploy large-scale GPU computing systems at scale. (Krupa et al., 2021).

Being that GPUaaS is frequently provided as SaaS, you can concentrate on creating, developing, and implementing AI solutions for end customers. Additionally, GPUaaS may be used with a server model. CPU power is consumed in enormous amounts by computationally demanding tasks. By offloading some of this work to a GPU, GPUaaS allows you to increase performance output and save up resources.

Popular Cloud Machine Learning Platforms

These three well-known machine learning platforms are provided by top cloud service companies.

AWS Sage Maker

The completely managed machine learning (ML) service offered by Amazon is called Sage Maker. It makes it possible to develop and train machine learning models rapidly and then implement them straight into a production setting. These are some of AWS Sage Maker's primary features:

- **An integrated Jupyter authoring notebook instance**: makes data sources easily accessible for investigation and analysis. Server management is not necessary.

- **Common machine learning algorithms:** The service offers algorithms that are tuned to operate well in a distributed setting while dealing with large amounts of data.

- **Native support for custom algorithms and frameworks:** SageMaker provides adaptable dispersed training systems that may be customized for specific procedures.

- **Quick deployment:** The service enables you to swiftly deploy a model into a scalable and secure environment using SageMaker Studio or the SageMaker interface.

- **Pay per usage:** Training and hosting are billed by AWS SageMaker use minutes. There are no upfront obligations or minimum fees.

Azure Machine Learning

An online tool called Azure Machine Learning aids in managing and expediting the whole machine learning project lifecycle. Workflows may be used to train and deploy machine learning models, build your own model, or utilize a model like Pytorch or TensorFlow. You can monitor, retrain, and redeploy your models with its help. It also enables you to handle MLOps.

This service enables both individuals and groups to implement machine learning models in a safe and auditable production setting. It contains tools for integrating models into services and applications, automating and speeding up machine learning operations, and using tools supported by robust Azure Resource Manager APIs.

Google Cloud AutoML

Google Cloud's machine learning service is called AutoML. A deep understanding of machine learning is not necessary. Building on Google's ML capabilities, AutoML can assist you in developing unique ML models that are suited to your particular requirements. It enables you to include your models into webpages and applications. These are AutoML's salient characteristics:

- **Vertex AI**—creates a single client library, API, and user experience for both AI and AutoML platforms. It allows you to obtain forecasts, store and deploy models, and use AutoML training as well as custom training.

- **AutoML Tables**—enables a whole group to automatically create and implement machine learning (ML) models on structured data at a large scale.

- **Video Intelligence**—This tool offers a number of ways to incorporate ML video intelligence models into apps and websites.

- **AutoML Natural Language**—This tool analyses the structure and meaning of documents using machine learning (ML), enabling you to build a custom ML model to extract data, categorise papers, and comprehend the sentiments of writers.

- **AutoML Vision**—facilitates the training of machine learning models to categorise photos using your own unique labels.

Cloud platforms remain to be the key in the progression of both ML and DL because tools that are made available in cloud platforms help in the development, deployment, and management of model. Here are some of the features that it is possible to leverage: automatic model retraining, hyperparameter optimization, and data versioning to name a few; in this case, cloud platforms give data scientists and developers more time to create value instead of worrying about infrastructure. Furthermore, such platforms make readily available pre-trained models and frameworks including TensorFlow, PyTorch, and Scikit-Learn that enable businesses to deploy AI solutions expeditiously and avert time delays common with AI implementation.

Also, integration of both ML and DL in cloud platforms also supports cooperation and elasticity, enabling teams to cooperate on projects yet have flexible resources. Organisations that work with big data or that need real-time processing for tasks such as fraud prevention, recommendation engines and diagnoses are especially benefited by this flexibility. That remains the case, and as cloud technologies advance, so too will improvements in ML and DL continue to expand their accessibility and expand the industries that leverage AI for better and more efficient decision-making.

2.4 Data Management for AI-Driven Cloud Solutions

Data management is a critical enabler of successful AI solutions that rely on cloud computing to organize and manage data for feeding into AI models for analysis. Cloud computing hosts unstructured as well as structured data within central hubs and this means that cloud data management must be done effectively. Good frameworks of data management address issues of data quality, security, availability, and conformity that are essential for AI efficiency. This helps to make the data derived from the analysis original, without bias and in line with the goal of AI solutions (Gupta & Sharma, 2023)traditional on premise solutions for data storage and analysis can become inadequate to handle the increasing volume, variety and velocity of healthcare data. The study aims to investigate the potential benefits and challenges of using cloud-based solutions for data analytics in healthcare. This paper reports about latest development and detailed role of using Artificial intelligence and capabilities of cloud Computing in health care sector/industry to foster innovative thinking, optimum wellbeing of the patient, focused medicinal support. This paper discusses various applications, algorithms and future of big data analytics with a focus on architecture, application and applicability of big data analytics using Hadoop and Cloud Computing in healthcare industry such as monitoring, prediction, performance, management etc including intensive care unit. many cloud platforms, like MMAP, are working in this field to provide a fast, reliable cost effective, efficient, and patient centric and solution to community health issues with capability of forecasting the health impact of various diseases on community for a given region or nation. Cloud computing framework, along with Artificial intelligence and Hadoop, aids healthcare management in completing analytical computations to identify logical, pertinent, and factual trends essential to strategize and enhanced readiness in event of catastrophes by facilitating data exchange among all stake holders. .","author":[{"dropping-particle":"","family":"Gupta","given":"Urvashi","non-dropping-particle":"","parse-names":false,"suffix":""},{"dropping-particle":"","family":"Sharma","given":"Rohit","non-dropping-particle":"","parse-names":false,"suffix":""}],"container-title":"2023 6th International Conference on Information Systems and Computer Networks, ISCON 2023","id":"ITEM-1","issued":{"date-parts":[["2023"]]},"title":"A Study of Cloud Based Solution for Data Analytics in Healthcare","type":"paper-conference"},"uris":["http://www.mendeley.com/documents/?uuid=9d76384d-08bc-47f1-903c-6e4e77cb0b2d"]}],"mendeley":{"formattedCitation":"(Gupta & Sharma, 2023.

The ability to store massive datasets from different sources is one of the critical challenges of data management in AI-consumed cloud environments by using storage solutions such as data lakes. AWS, Azure and Google cloud platform offer real-time data ingestion, transformation, integration tools that are crucial for constant AI model improvement. Data lifecycle and data orchestration guarantee continual freshness of data meaning AI models are trained on the most current data. Further, it is necessary to perform metadata management, lineage tracking, and versioning of data to maintain data quality across the AI life cycle.

However, it is a basic necessity to integrate security and compliance in AI based cloud systems. With the rising complexity of the data protection regulations, like GDPR and CCPA, data encryption, access control, or anonymization became crucial to data protection. Other tools in the form of AI can add to this process by automating data classification, risk assessment, and security enforcement. Going forward, the focus will be on further dynamics of the AI and the role that effective data management frameworks play in achieving the full potential of the AI in different sectors with the maximum minimization of risks and with the maximum adherence to the ethical use of the data.

Data governance is a fundamental component of AI-cloud solutions and refers to an organization's ability to acquire, store, process, and apply data to the purpose of enabling enhanced analytics and machine learning techniques. The availability of data from multiple sources including the Internet of Things, social networking sites, business applications, etc., has been increasing at a very rapid pace, thus requiring efficient techniques of data management. The cloud platforms have better features such as distributed storage and data pipelines, data lakes which is necessary when dealing with big and heterogeneous data for artificial intelligence workloads. Good data management means that AI models get the best quality data to produce better prediction and analysis and minimizes problem like data bias and data inconsistency.

Both scalability and flexibility in terms of data storage are important for AI employing cloud computing. The cloud hosting providers AWS, Microsoft Azure, and Google provide petabyte-level solutions for structured and unstructured data storage, such as Amazon S3, Azure Blob Storage, and Google Cloud Storage, respectively. Since they offer a focal point for many data kinds that machine learning algorithms may immediately access, the ideas of data lakes and warehouses are generally applicable. Additional technologies, including Apache Kafka and AWS Kinesis, are available to facilitate data ingestion and processing for continuous data as well as real-time AI computation.

Data preparation and data transformation are two critical steps in the management of AI data. Original data from various sources tend to have problems like missing values, errors that are disruptive to the functioning of AI systems. Other cloud-based data transformation tools are Databricks and AWS Glue through which data cleaning, normalization and feature engineering can be done. Another benefit of real-time data pipelines is that they guarantee that the data used to train AI models is up-to-date, reducing the likelihood of wrong data being perpetuated.(Johnson et al., 2024).

Both security and data management are equally important in AI-based cloud platforms. Since organizations come across personal information (PII) and financial data, cloud solutions implement cryptographic methods, roles, and compliance solutions. Additional security precautions adopted in the cloud include the use of multi-factor authentication (MFA), restricted role-based access controls (RBAC), and data encryption both in transit and at rest. Certain laws, such as the CCPA, GDPR, and HIPAA, require businesses to implement data security safeguards, which cloud platforms facilitate with integrated compliance tools and trail logs.

In addition, metadata management plays a critical role of maintaining data integrity as well as data lineage tracking of the data passed through AI systems. Metadata tools enable organisations to discover where the data has come from, how it has been processed along AI life cycle and how it is used. It is not only the compliance way, but also the collaboration of data scientists, engineers, and business stakeholders have a clear perspective of how data passes through the system.

Therefore, greater insight into data management in AI driven cloud solutions would require the consideration of data storage, pre-processing, security and governance. Through the adoption of the cloud platforms, it is possible for organizations to improve on the quality of data feeding their AI models, and enhance security in the data feeding the models to produce better insights, decision-making, and ideas. Basically, good data management practices will continue to be core competencies as AI innovations persist in the coming years.

2.5 Natural Language Processing in Cloud-Based Applications

A new cloud feature that is rapidly developing is Natural Language Processing (NLP) which allows their machines to comprehend, analyse, and even compose human speech and text with high precision. Growing importance of analytics in driving organizational decision making also puts heavier reliance on NLP for tasks like text mining, language translation, sentiment analysis, and even conversational interfaces. The availability of sophisticated NLP technologies in the cloud means that companies can achieve improved results from the tools without requiring massive on-premise infrastructure. These capabilities have made it possible to automate customer support, real-time translation, content moderation, and market sentiment analysis amongst other things, improving both customer experience and organizational effectiveness across sectors (Rayhan et al., 2023).

As leaders in cloud-based natural language processing (NLP), Google Cloud, Microsoft Azure, and Amazon Web Services (AWS) include them into their platforms. Google Cloud Natural Language AI models that can extract entities and feelings and classify text using machine learning methods are now available for usage. Same as that, we have Amazon Comprehend is an NLP service provided by AWS which pre-processes data by identifying key phrases, sentiment and language of unstructured text making it easier to analyze documents or customer feedbacks. Under Azure Cognitive Services, Microsoft Azure contains a number of NLP tools such as text analytics and language understanding (LUIS) that allow developers to create intelligent apps even without knowledge in data analysis.

The integration of NLP services into the clouds is pertinent to multiple advantages. These services are very flexible in terms of volume of data, this is because organizations dealing with these services can be prepared to deal with large volumes of data with little or no delay. Moreover, cloud-based NLP models are updated by their service providers and receive routine updates to the language models and the service. It reduces the entry barriers for companies that are small or start-ups to adopt sophisticated language technologies in their operations, and that too with the aid of minimal resources. While the countries and industries go on with the process of digitalization, the cloud-based NLP solutions will remain business-critical for such activities as marketing with personal approach, fraud identification, knowledge processing, and immediate customer interaction.

Artificial Intelligence, specifically Natural Language Processing (NLP) is revolutionizing ways organizations and companies engage and analyze large volumes of text-based information. As one of the components of cloud applications, NLP, machines the ability to interpret, learn and interact with human language as people would do. NLP releases the vast text and voice-based datasets that make it easier for several industries such as healthcare, finance, retail, and customer service organizations to enhance their performances and services. For example, NLP is beneficial for better document organization and handling, improves chatbots and virtual assistants, and helps to produce better sentiment analysis and get additional insights to understand customers better. However, with the development of cloud technologies, the requirements and the complexity of NLP procedures have been minimized and made available to businesses of different sizes.

Another advantage of using NLP on clouds is this feature of scale and flexibility of the cloud platforms. Businesses may take use of the greatest NLP solutions that many cloud providers, like as Google Cloud, Microsoft Azure, and Amazon Web Services, have incorporated into their cloud platforms. Google Cloud, for example, Natural Language API provides services such as Entity recognition, Sentiment analysis, Syntax, and Text classification. Such tools have the capability to recognize and analyze different languages which makes it possible for massive organizations with customer from different parts of the world. Similarly, AWS's Amazon Comprehend delivers pre-built machine learning models that are sufficient to extract entities, relationships and key phrases from a large volume of textual data. Given their complexity and time-consuming, sometimes human error-prone analysis, these services allow organisations to extract value from massive amounts of data that are typically beyond controllable analysis.

While others have gone as far as offering individual APIs for specific services, Microsoft's Azure Cognitive Services it up a notch offering an entire package of language related services including the language understanding (LUIS), text analytics and translator services that can help businesses develop applications that can carry out real-time conversations. These NLP tools are enable developers create intelligent chatbots, voice assistants and sophisticated text analysis systems that are capable of capturing queries from users, processing such queries comprehensively and respond with an acceptable feedback. It not only makes the customer support much easier to manage but also allows the business to create more individual interactions with the customers. For instance, in e-commerce this application is utilized in enhancing recommendation systems, personalized communication, and even responding to customers' inquiries. Besides, the use of NLP into virtual assistant, home automation, and car information systems is improving customer experience through interaction with devices, making them natural and smart.

Cloud-based NLP solutions can also create business opportunities that take advantage of the continuously updated language models which can improve the accuracy of the system over time. These models are trained with huge dataset, this makes it possible to understand the variation between different languages, different dialects and even industry specific terminologies. Automatic update and enhancements from the cloud providers make it possible for the businesses to avoid buying and updating their language models. This makes NLP solutions more affordable, quicker and less complicated to deploy, which is beneficial for small businesses and start-ups which do not have a plethora of AI knowledge on board. In addition, the cloud infrastructure guarantees high availability of the NLP models, that can address increased traffic when the demand arises without the necessity of slowing down the models (Supriyono et al., 2024).

The other benefit for cloud adoption of NLP is the possibility of analysing big data sets in real time. This is particularly so in areas such as the financial sector where sentiment of articles and posts, company reports and tweets can still influence trading activities. Likewise, in healthcare, NLP helps to review patients' chart, literature and clinical notes, for diagnosis and treatment purposes. In retailing for instance, through textual data analysis, companies are able to analyse customer reviews, social media posts and other textual information in order to determine the public's perception, discover new trends, and make necessary enhancements to goods or services offered.

As more corporations advance towards the adoption of digital technologies, these cloud-based NLP solutions should also gain more prominence in determining the relationship that organizations have with data. From improving customers' experience, streamlining business processes, providing better search capabilities, and analytically mining texts, cloud applications based on NLP provide substantial operational benefits. With such a technology as AI, there is always room for improvement, and with the emerging new features like contextualization, and multilingualism, and better emotional intelligence, AI and cloud technologies are poised for the next level of development (Pais et al., 2022).

Thus, the utilization of NLP in cloud-based applications defines a new novel way of making sense and gaining value out of the unstructured data in the businesses' context. Using efficient and elastic cloud environments, OOD can easily obtain the state-of-art NLP solutions for such processes as sentiment analysis, entity recognition and language translation, without worrying about strong computational and language backgrounds in-house. The integration of NLP as a cloud-based application not only results in the optimisation of business procedures but also facilitates the continuous monitoring of data, hastening enhancement in the field of customer satisfaction besides making organisational processes and decisions much quicker across healthcare, financial, and retail sectors. In the future, NLP solutions will act as enablers of changes, which will help make the AI language understanding more accessible and useful across different enterprises.

2.6 Cloud AI Tools and Frameworks

Cloud AI tools and frameworks act as ready-made platforms that organizations can effectively utilize to build and deploy Artificial Intelligence AI solutions at scale. These tools free you from the need to set up enormous on-premise facilities since the services they provide are cloud-based for constructing, training, and deploying AI models. Cloud AI platforms facilitate the adoption and application of AI by enterprises by providing a range of services, including model building, computer vision, natural language processing, and other predictive analytical models. Important cloud platforms like AWS, Microsoft Azure, and Google Cloud have developed robust AI stacks and customized solutions like AWS SageMaker, Azure Machine Learning, and Google AI Platform to meet this demand as AI engineering has swiftly become an industry standard. These platforms provide complete solutions, including data pre-processing, model building, implementation, and even monitoring, which provides complete advantages for AI development (Haefner et al., 2023).

One of the most important benefits of cloud-based AI tools is the compatibility with the most commonly used open-source frameworks, including TensorFlow, PyTorch, Keras, Scikit-learn, and MXNet. This integration enables data scientists and developers to use tools that are already well known to them, but with the added benefits of the cloud computing technology. Most platforms also have libraries of pre-built models and APIs for tasks like image classification, speech processing, and translation which can be easily integrated into an application without needing domain-specific knowledge on model creation. AutoML which is now becoming a common feature in AI platforms provides an even easier way to use AI since it removes the drudgery work of feature engineering, model selection and hyperparameter tuning that are usually time-consuming and which can only be undertaken by data scientists and analysts (N. L. Rane et al., n.d.).

In addition to concept development, cloud AI frameworks give importance to interaction and efficient data acquisition. Most of these tools integrate version control, data, and collaboration tools to ensure that multiple teams can collaborate well on AI projects successfully. Adapters with data lake, storage service and ETL (Extract, Transform, Load) help to facilitate data transfer across the AI pipeline stages, from data input to model deployment. In addition, these platforms are not limited to a cloud environment and are can also operate in a multi-cloud and hybrid cloud, meaning businesses can adapt their AI operations to the different environments (Chowdhury et al., 2022).

With the constant advancement of cloud AI tools, these are now key in the dissemination of AI in different sectors. The availability of the platform also increases through the advancement of low-code and no-code methods for the development of AI models with simple graphical user interfaces. Moreover, due to the pay-as-you-go model, cloud services do not require any large investments, which means that AI can be successfully integrated into start-ups and offer big businesses. As for the future development, cloud-based AI tools and frameworks will further improve by the emerging of federated learning, edge computing and real-time AI will enhance the innovation and operation efficiency among industries.

2.7 Chapter Summary

In summery focused on unveiling the combination of Artificial Intelligence (AI) and cloud computing showing how cloud solutions offer elastic resources and sophisticated utilities for AI solutions. It showcased the major AIaaS, machine learning, deep learning frameworks and NLP tools that are available in cloud environments. The chapter also covered data management of AI-based cloud applications, including data storage security, data flow, and privacy. Moreover, it discussed about AI as a service tools and platforms like AWS Sage Maker, Google AI Platform and Azure Machine learning how they are helping businesses to automate their AI model building and deployment process. When AI meets the cloud, organizations will be able to introduce innovation, superiority in automation, and efficiency in any decision-making mechanism.

Multiple Choice Questions (MCQs)

1. **What is the primary purpose of AI as a Service (AIaaS)?**

   a. To provide physical AI hardware to clients

   b. To offer AI capabilities through cloud platforms without infrastructure management

   c. To sell AI-powered gadgets for home use

   d. To replace traditional cloud services with AI models

2. **Which of the following best describes Machine Learning in cloud platforms?**

   a. A method to store large datasets in the cloud

   b. A service that provides pre-built AI models for developers

   c. The automation of all cloud resources without human intervention

   d. The process of manually analysing data without automation

3. **What role does data management play in AI-driven cloud solutions?**

    a. It restricts the amount of data processed by the AI model

    b. It ensures the quality, security, and availability of data for AI applications

    c. It reduces the need for cloud storage

    d. It focuses only on deleting redundant datasets

4. **Which of these technologies enables machines to understand and respond to human language?**

    a. Deep Learning

    b. Reinforcement Learning

    c. Natural Language Processing (NLP)

    d. Computer Vision

5. **Which cloud AI tool is commonly used for building and deploying machine learning models?**

    a. Google Cloud AutoML

    b. Adobe Photoshop

    c. Microsoft Word

    d. AWS Lambda

6. **What is a key benefit of using cloud platforms for AI workloads?**

    a. Reduced need for data encryption

    b. Limited scalability for AI models

    c. On-demand computational resources for AI model training

    d. Manual resource management for all AI models

7. **How does Deep Learning differ from Machine Learning in cloud environments?**

    a. Deep Learning requires manual feature extraction, while Machine Learning is automated

    b. Deep Learning uses neural networks to handle complex datasets, while Machine Learning relies on simpler algorithms

    c. Machine Learning is only used for small datasets, while Deep Learning requires no data

    d. Both are identical in how they process data

8. **Which of the following is a commonly used framework for developing cloud-based AI models?**

    a. TensorFlow

    b. GitHub

    c. Zoom

    d. Windows Defender

9. **What is a major challenge in integrating AI into cloud platforms?**

    a. Lack of scalability

    b. Difficulty in securing large datasets

    c. Limited AI tool availability

    d. Inability to handle complex data processing

10. **What does Natural Language Processing (NLP) mainly focus on in cloud-based applications?**

    a. Enhancing image quality

    b. Analysing and interpreting text and speech data

    c. Providing secure cloud storage

    d. Accelerating data transfer speeds

Answer

1	2	3	4	5	6	7	8	9	10
b	b	b	c	a	c	b	a	b	b

Chapter 03

AUTOMATION IN THE CLOUD ERA

3.1 Chapter Overview

This chapter introduces chatbots and virtual assistants in automated cloud services. Their changes to customer assistance, marketing, and content generation are listed. I emphasise the benefits of AI-based chatbots: rapid answers, infinite queries, cost savings, individual approach, translation availability, and 24/7 availability. These tools simplify work, reduce errors, and boost client happiness. Language, context, data privacy, and implementation challenges are also discussed in the chapter. Chatbots can improve customer experience and automate work processes by answering frequently asked questions, booking appointments, suggesting and upselling products, placing orders and tickets, and escalating enquiries to technical support.

3.2 Automation in Cloud Contexts

A wide range of procedures and instruments that lessen or do away with the need for human intervention in the provisioning and administration of cloud computing workloads and services are together referred to as cloud automation. Cloud automation may be implemented in private, public, and hybrid cloud settings by organisations. (Josyula et al., 2011).

Why Use Cloud Automation?

Enterprise workload deployment and operation are traditionally done by hand, which takes a lot of time. This frequently entails monotonous duties like the following:

1. Resource sizing, provisioning, and configuration, including virtual machines (VMs).

2. Setting up load balancing and virtual machine clusters

3. generating logical unit numbers (LUNs) for storage.

4. Using virtual networks.

5. The deployment of the cloud itself.

6. Keeping an eye on and controlling performance and availability.

Even while all of these manual and repetitive procedures work, they are ineffective and frequently prone to mistakes. The availability of the workload may be delayed as a result of troubleshooting caused by these problems. These could potentially reveal security flaws that could endanger the company.

An organisation can get rid of these manual and repetitive workload deployment and management procedures by implementing cloud automation. An IT team must employ orchestration and automation solutions that operate on top of its virtualised environment in order to accomplish cloud automation.

Types of Cloud Automation

Cloud automation eliminates repetition, inefficiency, and errors from manual procedures and interventions. These are typical examples:

- **Allocation of resources:** Cloud computing relies on autoscaling to match demand for processing, memory, and networking resources. It provides resource elasticity and the pay-as-you-go cloud cost model.

- **Configurations:** Infrastructure setups that are represented by code and templates can be implemented via automation. As related cloud services are added, cloud integration grows.

- **Development and deployment:** Continuous software development automates code scans, version control, testing, and deployment.

- **Tagging:** Context, operation, and criteria are the foundations of automatic asset tagging.

- **Security:** Automated security controls can check cloud environments for vulnerabilities and unusual performance, as well as authorize or restrict access to apps or data.

- **Monitoring and logging:** Every service and workload activity can be recorded by cloud tools and functionalities. Monitoring filters are able to identify surprises and anomalies.

Benefits of Cloud Automation

The following are just a few advantages of cloud automation when done correctly:

- Saves a company money and time.

- Is more scalable, safer, and quicker than carrying out activities by hand.

- Reduces error rates by enabling organisations to create more dependable and predictable workflows.

- Increases efficiency by enabling continuous deployment and automating bug detection.

- Simplifies implementation, compared to on-premises platforms, requiring less IT intervention.

- Directly supports improved corporate governance and information technology.

- Enables IT staff to concentrate on higher-level work that better fits the business demands of the company by relieving them of manual and repetitive administrative duties. This involves creating new product features or using more advanced cloud services.

Challenges of Cloud Automation

Cloud automation can also have the following challenges:

- Internet connectivity cannot be guaranteed. The reliability of public cloud services' wide area networks should be discussed with the supplier.

- Limited security options in cloud automation might be challenging in highly regulated businesses with complicated compliance requirements due to limited customisation and control flexibility.

- Limited access to back-end data can pose maintenance challenges for complex issues.

- Platform lock-in poses a concern. Cloud automation's simplicity can encourage enterprise adoption, with more business processes and operations adopting the platform. The larger the commitment, the harder it will be to switch platforms.

3.3 DevOps and Continuous Integration/Continuous Deployment (CI/CD)

DevOps emphasises collaboration between developers and operations workers to minimise the discrepancies between developing and running and increase value-proper delivery by automating the software delivery process. DevOps' CI/CD technique automates the integrated change and code delivery process of small code snippets being merged and published to production environments quickly (Erich et al., 2017).

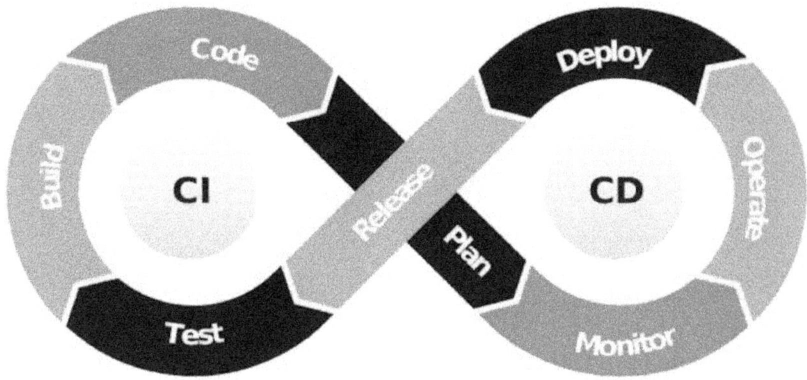

Figure 3.1: The image you sent is a visual representation of the DevOps Lifecycle.

Source: - *(Tecxar, 2022)*

CI/CD

The objective of Continuous Integration (CI) and Continuous Delivery (CD) is to create a pipeline of development, testing, and deployment. CI/CD lets developers work on code simultaneously with minimum overhead. Because each programmer uses the latest project codebase, the end output has fewer errors.

Benefits of CI/CD

- Automated tests run continuously (CI) or after every commit, reducing manual testing time.

- Developers operate on their own repository, reducing conflicts and avoiding "mine" vs "yours."

- Automation tools enable faster deployments by executing scripts based on build conditions.

- Enhanced software quality

CI/CD helps DevOps improve software quality. This allows you to automate testing for every code build, ensuring product quality.

A continuous integration system can also discover integration or compilation issues, allowing teams to repair flaws without manual testing at the conclusion of a project cycle. This method lets engineers resolve flaws quickly rather to waiting until completion, speeding up software development.

Increased software development speed

CI/CD speeds up development by allowing you to run tests continuously on a live system rather than manually in separate environments and wait until completion before moving on to another task or stage, which is a huge time-saver in Agile sprints, where multiple people work on different parts at once while still integrating everything.

Implement CI/CD

CI/CD implementation in your organisation requires multiple steps. The first stage is to establish a pipeline, which transfers code from development to testing to production. A simple pipeline diagram is shown below:

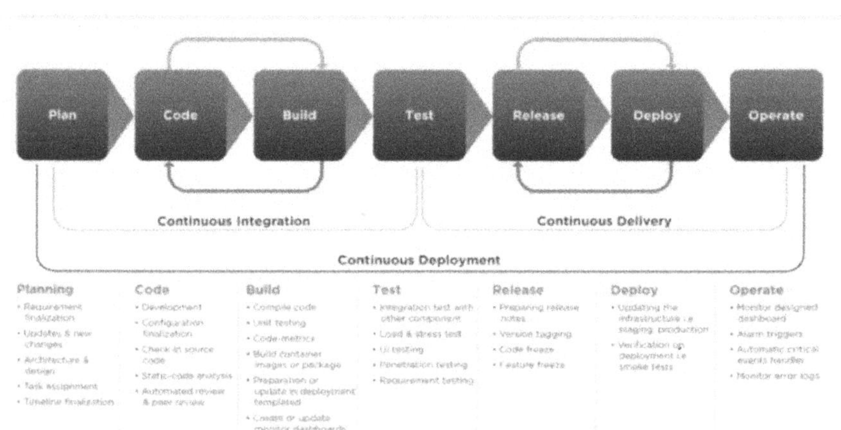

Figure 3.2: The image you sent is a visual representation of the DevOps Lifecycle. It shows the continuous flow of software development and delivery, starting from planning and ending with monitoring. The key stages are Plan, Code, Build, Test, Release, Deploy, and Operate.

Source: - *(Tecxar, 2022)*

Tools for implementing CI/CD.

- CI/CD is a process that unites several entities to streamline software development, rather than just a toolkit. CI/CD tools support this approach, although they don't have to be used together.

- Jenkins simplifies CI/CD implementation for teams of all sizes and skill levels with powerful automation tools. Its plug-in support and vast community make it a popular choice among DevOps specialists.

- Bamboo is a popular alternative for establishing CI/CD techniques, allowing for centralised repository management using a dashboard interface. This solution interfaces with many DevOps ecosystem technologies and provides good reporting to help IT teams who are just starting out with automated deployment processes meet compliance standards.

Continuous Integration

A team of developers may spend a lot of time working independently before merging their changes into the main branch. Because of this, code merging becomes difficult, time-consuming, and prone to conflicts. It also results in long-term problems that are discovered later in the development process. Customer updates are slowed by these causes. During Continuous Integration, developers commit to a shared repository using Git. Sonar can be used for code review, unit testing, build execution, and artefact storage in an automated continuous integration pipeline. Our CI workflow may be triggered by every codebase commit/merge.

Continuous Delivery

Developers can avoid last-minute surprises by testing their code in a production-like environment with continuous delivery. These tests may include user interface, load, and integration tests. It helps developers find flaws early on and fix them.

The software release process is automated by CD, which leads to low-risk releases, lower costs, better software quality, more productivity, and—

above all—faster and more frequent customer updates. If Continuous Delivery is successfully applied, we will always have tested code that is ready for deployment.

We use Continuous Delivery (CD) following Continuous Integration to distribute new updates to clients rapidly and error-free. The final release won't disrupt production if integration and regression tests are run in staging, which is identical to production. By automating the release process, we can deploy our app and ensure that our product is always ready for release. Software releases are automated by continuous delivery. The project manager or developer can deploy to live production as needed. AWS Code Deploy, Jenkins, and GitLab are a few well-known CD tools.

Continuous deployment

CI and CD end with continuous deployment. Continuous delivery, which automatically adds the appropriate code to the code repository, is extended by continuous deployment. Because there is no manual gate before production, continuous deployment automates the related app. The developer's cloud-based application change gets live within minutes of writing pass automated testing.

The CI workflow

A CI pipeline is depicted in this graphic, which starts with developers checking in code and ends with automated build, test, and build status notifications.

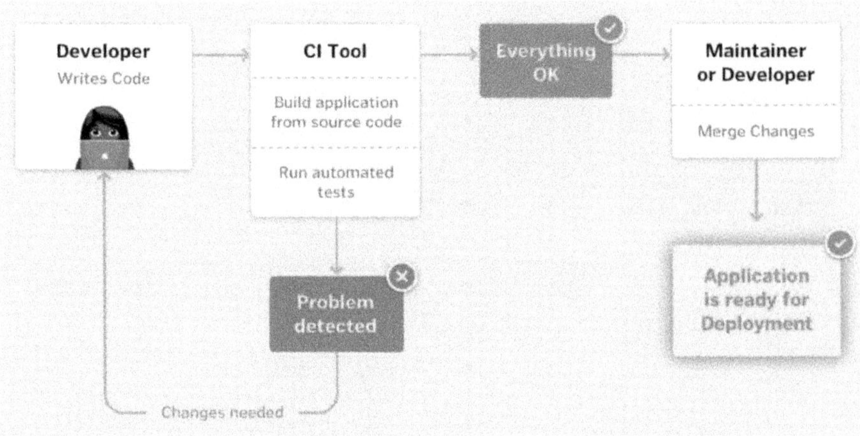

Figure 3.3: The image illustrates the Continuous Integration (CI) process. A developer writes code, which is then built and automatically tested by a CI tool. If the tests pass, the changes are merged, and the application is ready for deployment. If tests fail, the developer is notified to fix the issue.

Sourse: - *(Greeksforgreeks, 2023)*

The CI pipeline downloads the changes and runs automatic build and unit tests after a developer commits code to Git. After integrating the new code, the server notifies the developer if the step was successful or unsuccessful.

This expedites the identification and resolution of issues, increases developer productivity by doing away with manual tasks, and expedites client updates. Integration of the development cycle has been shown to save developer time by 25–30%.

CI and CD Workflow

The graphic below illustrates how Continuous Integration and Continuous development work together to accelerate software development while lowering risks and enhancing quality.

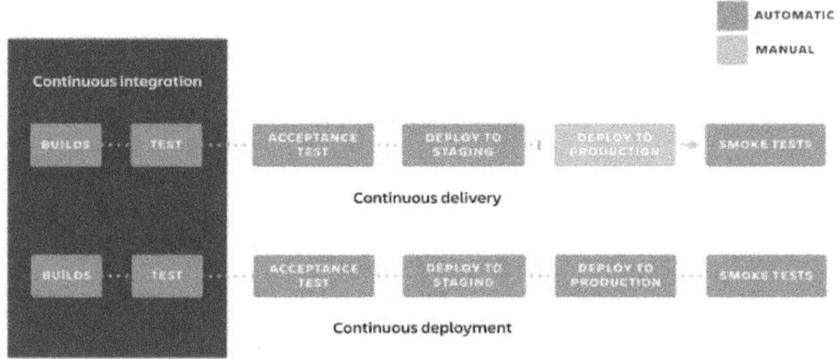

Figure 3.4: The distinctions between continuous deployment, continuous delivery, and continuous integration are depicted in the graphic. CI focuses on automating builds and tests. CD automates the release process up to the point of manual deployment to production. Automating the entire process, including deployment to production, is what continuous deployment does.

Sourse: - *(Greeksforgreeks, 2023)*

Building, testing, and packaging source code is automated via continuous integration after developers add it to the code repository. After CI, the code is deployed to staging for automated testing (Acceptance, Regression, etc.). Final product release occurs when it is deployed to production.

If production deployment is manual. That's called Continuous Delivery, while automatic production deployment is termed Continuous Deployment.

CI/CD Important

CI/CD makes it possible for businesses to create software fast and effectively. CI/CD makes it possible to move code into production constantly, ensure a constant flow of new features, and deliver software and goods to market more quickly than ever before.

Some Common CI/CD Tools

Teams may use CI and CD technologies to help with testing, deployment, and development. It is advised to use some for integration and others for testing and related tasks. The most widely used CI/CD tool is Jenkins. It is open source, capable of managing all kinds of tasks, and capable of creating a simple Ci server for the CD hub.

There are numerous other sources for handling CI and CD besides Jenkins:

- Concourse: An open-source tool for development of CI and CD mechanisms.

- GoCD is used for modelling and visualisation.

- Screwdriver is a CD-building platform.

- Spinnaker: a CD platform for multi-cloud environments.

Continuous integration vs. delivery vs. deployment

Aspect	Continuous Integration (CI)	Continuous Delivery (CD)	Continuous Deployment (CD)
Definition	Automating the process of merging code changes into a shared repository frequently.	Automating the release process so the software is always ready to be deployed.	Fully automating the release process, deploying changes directly to production.
Focus Area	Code integration and validation.	Ensuring deployable software at any time.	Automating deployment to production environments.
Key Process	- Frequent code commits. Automated builds. - Automated testing (unit tests).	- All CI processes. - Automated integration tests. - Manual approval before production deployment.	- All CI and CD processes. - Automated deployment to production without manual intervention.

Aspect	Continuous Integration (CI)	Continuous Delivery (CD)	Continuous Deployment (CD)
Deployment Frequency	Not applicable (focus is on code commits).	Deployment occurs manually or semi-automatically.	Deployment is continuous and automatic.
Manual Intervention	Required for subsequent stages (e.g., delivery or deployment).	Required to initiate production deployment.	None—fully automated.
Rollback Complexity	Not involved directly in rollbacks.	Rollbacks involve testing deployable versions.	Rollbacks must be automated or use feature flags for safe changes.
Primary Tools	Jenkins, Travis CI, GitHub Actions.	Jenkins, CircleCI, GitLab CI/CD.	Spinnaker, Octopus Deploy, AWS Code Deploy.

Sourse: - *(Greeksforgreeks, 2023)*

3.4 Infrastructure as Code (IaC): Simplifying Cloud Management

IT infrastructure management and provisioning using code is called Infrastructure as Code (IaC). Teams may automate infrastructure setup and administration for efficiency and consistency. This is helpful in DevOps, where teams regularly update and deliver software.

Using Infrastructure as Code

Infrastructure as Code helps you define and manage system infrastructure in code. IaC describes the system architecture, including servers, networks, storage, and operating systems, just like software code describes application logic and functionality.

IaC organises infrastructure resource management. Version control, error tracking, and easy upgrades are feasible because they consider configuration files like source code. Config files can be authored in Python or Java and developed in IDEs to spot mistakes and optimise coding. Version control systems track changes and let developers collaborate.

Role of IaC in DevOps

IaC's function in DevOps is to coordinate development and operations teams to produce software more quickly and with higher quality. Infrastructure as Code facilitates the automation of pipelines for continuous delivery, continuous integration, and infrastructure management.

DevOps might automate infrastructure changes based on application changes with IaC. No manual error or change is done similarly across environments.

Features of IaC

- **Automation:** IAC reduces human error and saves time by automating infrastructure provisioning and configuration.

- **Repeatability:** The reusability of IAC scripts makes it simple to replicate the same infrastructure across several contexts.

- **Version Control:** IAC code is kept in version control systems, like as Git, which facilitate collaboration, tracking of changes, and reversion to earlier versions.

- **Scalability:** Scaling infrastructure up or down and adding or deleting resources as needed is simple with IAC.

- **Transparency:** IAC provides transparency and comprehensibility to the infrastructure by defining its components and their interrelationships through the code.

- **Improved Security:** IAC lowers the risk of security flaws by assisting in ensuring that infrastructure is configured consistently and safely.

Applications of IaC

Applications for Infrastructure as Code are many and span several fields, including

- Cloud computing. Databases, storage, and virtual machines are among the cloud services that are frequently provided and configured using IAC.

- IAC plays a crucial role in DevOps, automating infrastructure and application deployment and management.

- CI/CD pipelines employ IAC to automate infrastructure and application deployment and configuration.

- IAC chooses a web server, application server, and load balancer automatically, automating the deployment and administration of web applications.

Advantages of IaC

- **Improved Reliability:** IAC ensures consistent, repeatable infrastructure, reducing manual errors and increasing uptime.

- **Faster Deployment:** IAC automates manual operations, enabling faster infrastructure and application deployment.

- **Enhanced Collaboration:** IAC facilitates information sharing and collaboration among multiple team members on infrastructure projects.

- **Improved Security:** IAC ensures consistent and safe infrastructure configuration, reducing security vulnerabilities.

- **Easier to Manage:** The IAC code simplifies infrastructure management by defining components and their interactions.

Disadvantages of IaC

- IAC involves expertise of scripting languages and cloud computing, resulting in a learning curve.

- To implement IAC, it takes time and effort to write scripts, test them, and integrate them into the existing system.

- IAC might be complex because of the need for various components to function together and the difficulty in debugging issues.

- IAC can build dependencies between components, making modifications or updates more challenging.

- IAC scripts can be vulnerable to coding errors, which can significantly damage infrastructure.

Use Cases of IaC

- IAC allows for the provisioning of Virtual Machines (VMs) in cloud computing environments, specifying the quantity, operating system, and required software.

- IAC is used for network deployment, topology design, subnet creation, and security group configuration.

- IAC can be used for load balancing, web server selection, application server configuration, and web application deployment.

- IAC can automate DNS record creation and deletion, guaranteeing uniformity across a variety of scenarios.

Popular Infrastructure as Code (IaC) Tools

Here's a list of some of the most commonly used Infrastructure as Code (IaC) tools, each with unique features and specific use cases:

Tool	Description	Approach	Key Features
Terraform	A widely-used open-source tool from HashiCorp that automates the management of cloud and on-premises resources.	Declarative	Multi-cloud compatibility, modular infrastructure, state management, extensibility.
AWS CloudFormation	An AWS-native tool for automating infrastructure deployment within the AWS environment.	Declarative	Deep integration with AWS, templated infrastructure setup, rollback capabilities.
Ansible	A simple, agentless automation tool for managing configuration, provisioning, and deployments.	Can be both Imperative and Declarative	Easy to use with YAML playbooks, large community, no agents needed.

Tool	Description	Approach	Key Features
Puppet	A configuration management tool designed for automation, with an emphasis on large-scale infrastructure.	Declarative	Manages complex setups, centralized server architecture, and strong community support.
Chef	A flexible tool that automates infrastructure management using a Ruby-based DSL (Domain-Specific Language).	Imperative	Supports advanced workflows and testing, with a focus on flexibility.
Salt Stack	A powerful automation tool for configuration management and orchestration, known for speed and scalability.	Both Imperative and Declarative	Real-time automation, event-driven, and security-focused.
Kubernetes	A platform for automating containerized applications' deployment, scaling, and management.	Declarative	Manages container lifecycle, networking, and scaling in a cluster environment.

Source: - (*geeksforgeeks, 2024*)

Declarative vs Imperative Approaches to IaC

Aspect	Declarative Approach	Imperative Approach
Definition	Specifies the desired state of the infrastructure, such as the resources and their configurations.	Details the exact steps or commands required to achieve the desired infrastructure state.
Execution	The IaC tool determines and performs actions to achieve the desired state automatically.	Requires the user to execute commands in the correct sequence to configure the infrastructure.
State Management	Tracks the current state of the infrastructure, simplifying updates and resource teardown.	Does not inherently track state; the user is responsible for managing changes manually.
Ease of Use	Simplifies the process as users define *what* they want, leaving *how* it is achieved to the tool.	Demands detailed instructions, making the user responsible for defining *how* to achieve results.
Change Management	Automatically applies changes when the desired state is modified.	The user must figure out and apply the changes manually.

Aspect	Declarative Approach	Imperative Approach
Tool Preference	Commonly used by most IaC tools, as it allows for automation and efficiency.	Less common, but some tools or scenarios may require it for greater control over actions.
Example Use Case	Defining infrastructure as a code template and letting the tool provision resources.	Writing scripts to create or update infrastructure in a specific order.

Source: *- (geeksforgeeks, 2024)*

How Does Infrastructure as Code Work in DevOps?

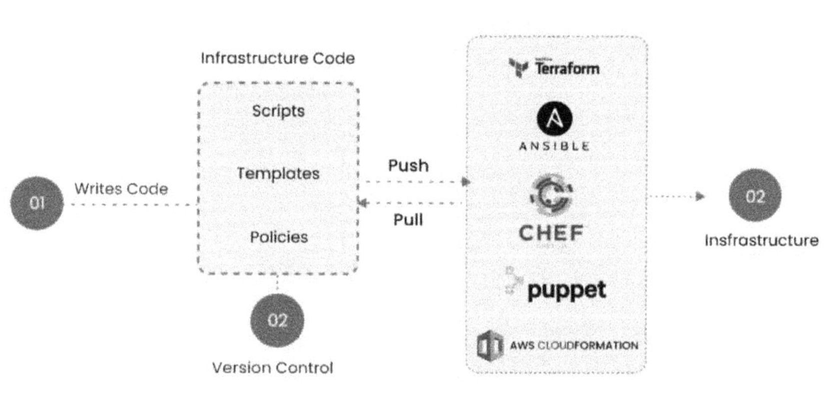

Figure 3.5: The figure illustrates the Infrastructure as Code (IaC) process. Developers write infrastructure code (scripts, templates, policies) and store it in version control. IaC tools like Terraform, Ansible, Chef, Puppet, and AWS CloudFormation then use this code to provision infrastructure

resources (like servers, networks, and storage) in a repeatable and automated manner.

Source: *- (Sinha, 2024)*

DevOps and DevSecOps use Infrastructure as Code to automate and manage infrastructure precisely. IT operations can be more efficient, scalable, and secure by treating infrastructure as code. Dig further into each stage:

Definition of Infrastructure: Teams code all infrastructure components, such as virtual machines, databases, and network settings. Teams describe these aspects in YAML, JSON, or HCL instead of manually. This helps developers define infrastructure requirements uniformly. Well-defined infrastructure through code allows environments to be reproduced across development, testing, and production without differences, maintaining consistency and eliminating errors.

Versioning: Version control is a major benefit of Infrastructure as Code. Git lets teams save infrastructure code in repositories. Every modification is tracked, so you know who did it, when, and why. Versioning lets teams revert to previous configurations if updates or deployments fail. This provides reliability and allows teams to make frequent changes without damaging infrastructure.

Automated Provisioning

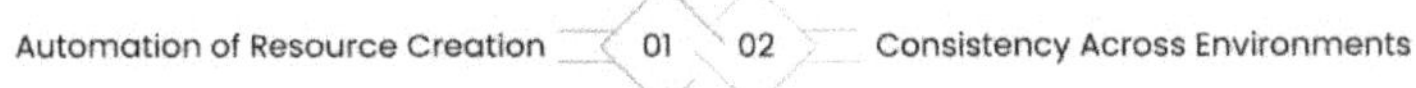

Figure 3.6: The image you sent represents the key benefits of Infrastructure as Code (IaC). IaC allows for Automation of Resource Creation by defining infrastructure resources (like servers and networks) in code, enabling automated provisioning and deployment. This also leads to Consistency Across Environments as the same code can be used to create identical

infrastructure in different environments (e.g., development, testing, and production).

Source: - *(Sinha, 2024)*

Terraform, CloudFormation, and Ansible are essential for infrastructure provisioning. The infrastructure code is read and resources are automatically created. Every stage of server launch, database setup, and network configuration is automated by these tools. Automating setup reduces errors and speeds deployment. Automated provisioning delivers infrastructure consistently across environments, ensuring reliability.

Benefits of IaC in DevOps

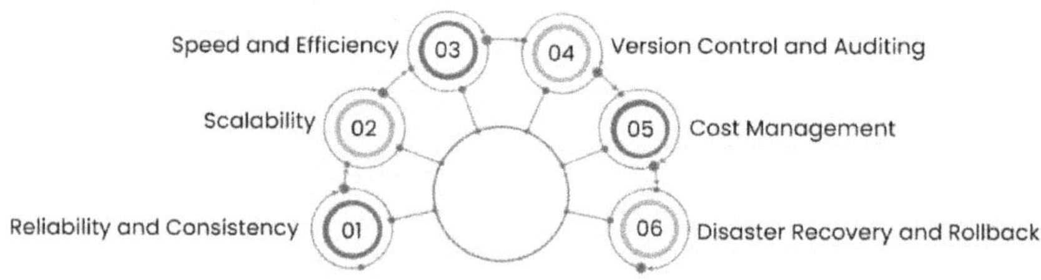

Figure 3.7: The figure highlights the key benefits of using Infrastructure as Code (IaC). These include increased speed and efficiency in provisioning infrastructure, improved scalability, enhanced reliability and consistency, better version control and auditing capabilities, reduced costs, and improved disaster recovery and rollback options.

Source: - *(Sinha, 2024)*

The Infrastructure as Code (IaC) model offers transformative benefits in the DevOps lifecycle. Let's explore these advantages further in detail:

- **Reliability and Consistency**

  In traditional IT environments, inconsistent configurations throughout development, testing, and production are a major issue. IaC writes infrastructure as code, ensuring consistent configuration. This avoids environment-specific issues and assures stage-wide consistency. It solves the "it works on my machine" problem, where a program works fine in development but not elsewhere. IaC standardises setups, reducing errors and improving system stability.

- **Scalability**

  Scaling infrastructure is easy with IaC. Scaling traditionally required manually adding servers, storage, and network configurations, which may take hours or days. This is done by upgrading IaC code. Scaling up for greater workloads or down for decreased demand can be done instantaneously. Terraform and AWS CloudFormation dynamically adapt infrastructure to code changes, allowing businesses to scale quickly without delays or errors.

- **Speed and Efficiency**

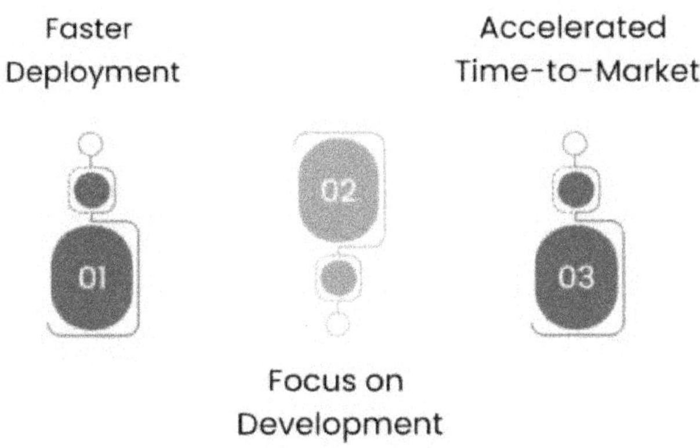

Figure 3.8: The image shows the benefits of using a platform like Google Cloud to accelerate your business. It highlights three key advantages: Faster Deployment, Accelerated Time-to-Market, and the ability to Focus on Development. By leveraging Google Cloud's infrastructure and services, businesses can streamline their operations, reduce time-to-market for new products and services, and free up their development teams to focus on innovation.

Source: *- (Sinha, 2024)*

In traditional IT environments, inconsistent configurations throughout development, testing, and production are a major issue. IaC writes infrastructure as code, ensuring consistent configuration. This avoids environment-specific issues and assures stage-wide consistency. It solves the "it works on my machine" problem, where a program works fine in development but not elsewhere. IaC standardises setups, reducing errors and improving system stability.

- **Scalability**

  Scaling infrastructure is easy with IaC. Scaling traditionally required manually adding servers, storage, and network configurations, which may take hours or days. This is done by upgrading IaC code. Scaling up for greater workloads or down for decreased demand can be done instantaneously. Terraform and AWS CloudFormation dynamically adapt infrastructure to code changes, allowing businesses to scale quickly without delays or errors.

- **Cost Management**

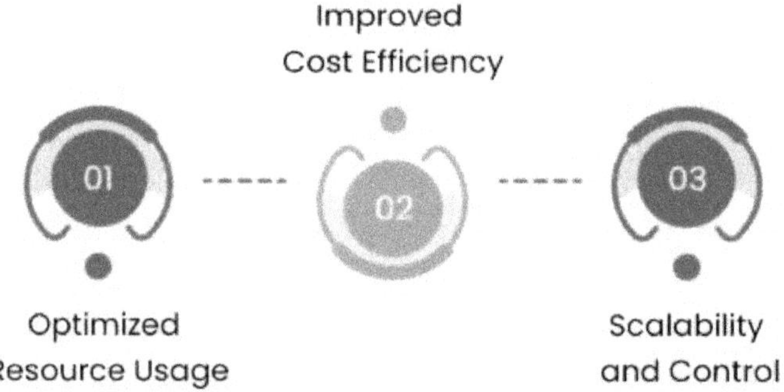

Figure 3.9: The figure illustrates the benefits of using a cloud platform like Google Cloud. It highlights three key advantages: Optimized Resource Usage, Scalability and Control, and Improved Cost Efficiency. By leveraging Google Cloud's infrastructure and services, businesses can

optimize their resource utilization, scale their operations as needed, and reduce their overall IT costs.

Source: - *(Sinha, 2024)*

Effective resource management is critical for cost control, especially in cloud computing. IaC allows organizations to create temporary environments for specific tasks, such as testing new features. Once testing is completed, these environments can be automatically de-provisioned to avoid incurring unnecessary costs. This level of flexibility ensures that resources are only utilized when necessary, optimizing cloud spending and preventing resource wastage. The ability to automate provisioning and decommissioning through code significantly enhances cost efficiency.

Overall, the infrastructure-as-code approach drives consistency, efficiency, and scalability across IT operations, making it indispensable for modern DevOps practices. It empowers teams to innovate faster while maintaining robust control over infrastructure, ensuring that businesses can scale seamlessly and respond quickly to changing demands.

Common Tools for Infrastructure as Code

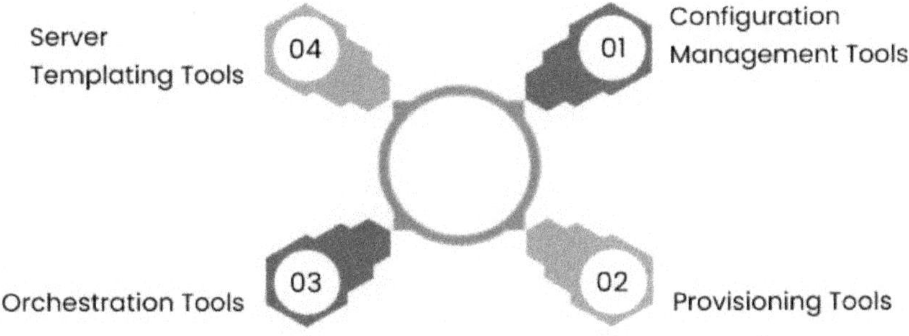

Figure 3.10: The image illustrates the different tools and technologies used in infrastructure management. These include Server Templating Tools, Configuration Management Tools, Provisioning Tools, and Orchestration

Tools. Each type of tool serves a specific purpose in automating and streamlining the management of infrastructure resources.

Source: - *(Sinha, 2024)*

DevOps team rely on four main types of Infrastructure as Code in DevOps tools to streamline operations. Each tool serves a distinct purpose in automating and managing infrastructure.

Configuration Management Tools

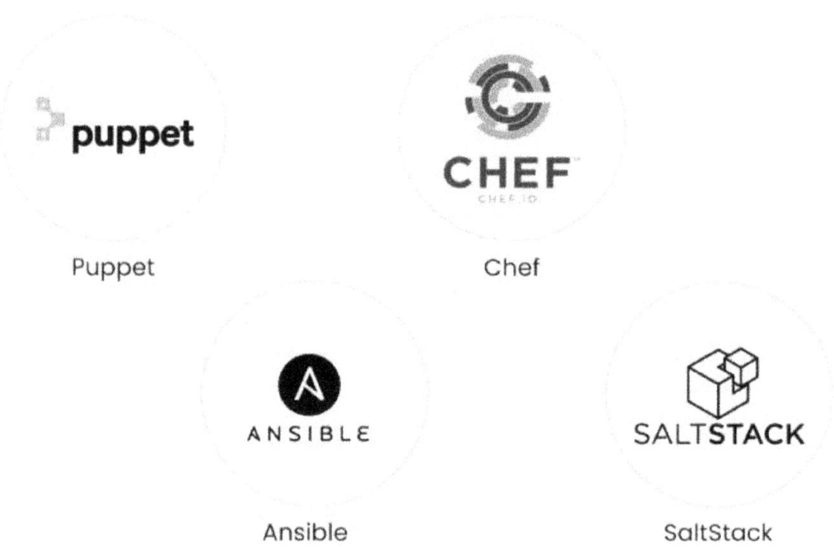

Figure 3.11: The image you sent shows the logos of four popular Infrastructure as Code (IaC) tools: Puppet, Chef, Ansible, and Salt Stack. These tools help automate and manage infrastructure resources like servers and networks by using code to define and provision them.

Source: - *(Sinha, 2024)*

Configuration management tools focus on automating software installation and managing configurations on existing servers. By ensuring uniform setups across all servers, they eliminate manual errors and configuration drifts.

1. **Puppet**

   Puppet shares many similarities with Chef, standing out among Infrastructure as Code in DevOps tools. It's integral to countless CI/CD pipelines that DevOps engineers rely on. Puppet uses a Domain Specific Language (DSL) based on Ruby, allowing you to declare the desired state of your infrastructure and its functionalities. It then automatically determines the most efficient path to achieve the specified configuration.

   If there are any deviations from this desired state, Puppet actively monitors and corrects these changes, restoring the infrastructure to its intended setup. This robust open-source solution facilitates smooth automation across a variety of contexts by supporting popular cloud platforms including Microsoft Azure, Amazon Web Services (AWS), and Google Cloud Platform (GCP).

Key Features:

- Puppet employs a Ruby DSL for defining system configurations, making scripting more intuitive and powerful.

- It offers an extensive module library through Puppet Forge, enhancing functionality and customization.

- Puppet's idempotent nature ensures that applying a configuration repeatedly yields consistent results, boosting stability.

2.   Chef

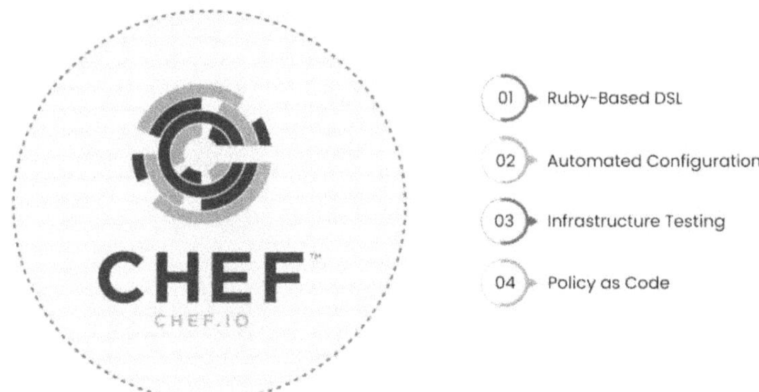

Figure 3.12: The image shows the Chef logo and highlights four key features of the Chef platform for infrastructure automation: Ruby-Based DSL (Domain-Specific Language) for defining infrastructure configurations, automated configuration management for consistent deployments, infrastructure testing for ensuring system stability, and policy as code for enforcing compliance and security rules.

Source: *- (Sinha, 2024)*

Chef is a powerful automation platform that converts Infrastructure as Code in DevOps into reality. It enables the automation of building, deploying, and managing IT infrastructure, providing organizations with precise control over configurations. This control drives greater agility and operational efficiency in IT processes, empowering teams to innovate rapidly.

Key Features:

- Chef utilizes a Ruby-based DSL to script system configurations, streamlining the automation process.

- It automated server configuration and deployment, reducing manual intervention and potential errors.

- Chef supports infrastructure testing with tools like Chef Spec and InSpec, ensuring reliable code quality.

- Policy as Code: It allows teams to define policies and configurations as code, enhancing transparency and consistency across deployments,

3. Ansible

Ansible serves as a versatile IT automation engine, ideal for application deployment, configuration management, and orchestration tasks. Although not a traditional Infrastructure as Code in DevOps tool, ANsible's adaptability allows it to provision IaC resources using various collections. This capability bridges the gap between IT automation and IaC, offering a unified approach to managing both software and infrastructure with a single toolset.

Key Features:

- Ansible operates without requiring agents on target nodes, simplifying the automation process.

- YAML-based automation: It leverages YAML to define tasks, making configurations straightforward and readable.

- Modular design: Ansible's modular structure supports reusable modules, enhancing flexibility for diverse automation scenarios.

4. Salt Stack

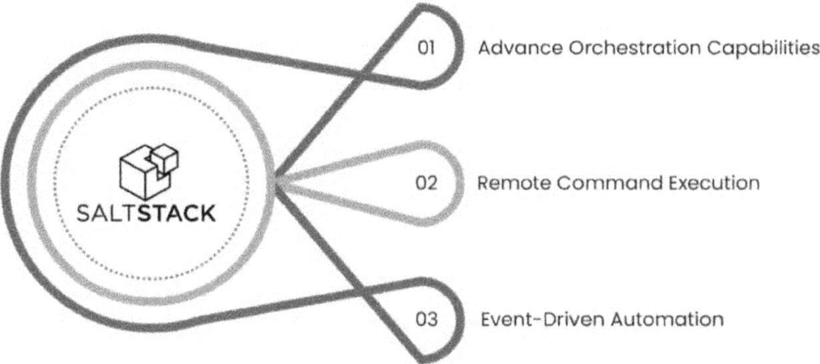

Figure 3.13: The image you sent shows the Salt Stack logo and highlights three key features of the Salt Stack platform for infrastructure automation:

Advanced Orchestration Capabilities for complex deployments, Remote Command Execution for managing and controlling remote systems, and Event-Driven Automation for reacting to real-time events across the infrastructure.

Source: *- (Sinha, 2024)*

Salt Stack, commonly referred to as Salt, is a robust Python-based tool specializing in Infrastructure as Code in DevOps. It's designed for efficient configuration management and remote command execution, excelling in both Cloud engineering and on-premises environments.

Key Features:

- Advance Orchestration Capabilities: Salt Stack supports intricate orchestration and configuration management across diverse environments, ensuring seamless infrastructure automation.

- Remote Command Execution: It allows direct command execution on remote systems, streamlining control over distributed infrastructure.

- Event-Driven Automation: Salt Stack reacts to various system events, triggering automated responses to ensure dynamic and responsive infrastructure management.

Server Templating Tools

Server templating tools allow teams to create reusable templates that define a server's configuration. These templates act as blueprints, standardizing how new servers are built and ensuring they meet predefined specifications.

Vagrant

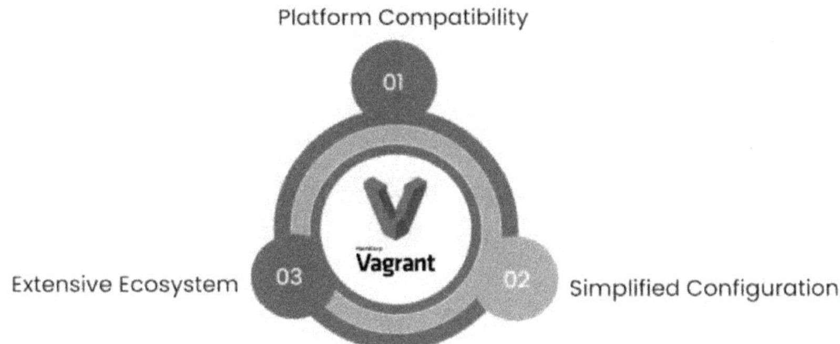

Figure 3.14: The figure shows the Vagrant logo and highlights three key features of the Vagrant platform for building and managing development environments: Platform Compatibility, Extensive Ecosystem, and Simplified Configuration.

Source: - *(Sinha, 2024)*

Vagrant streamlines the creation and management of virtual machines, enhancing Infrastructure as Code in DevOps practices by ensuring consistent environments. It allows DevOps engineers to easily share VM setups, promoting seamless collaboration and reducing environment-related issues.

Key Features:

- **Platform Compatibility:** Integrates with VirtualBox, VMware, Docker, AWS, and other major platforms for versatile virtual machine deployment.

- **Simplified Configuration:** Uses a straightforward Vagrant file to define and configure environments, making setup intuitive and user-friendly.

- **Extensive Ecosystem:** Offers a variety of preconfigured boxes for different operating systems, accelerating deployment with ready-to-use templates.

Orchestration Tools

Orchestration tools manage the coordination between different infrastructure components, ensuring seamless interactions in complex environments. They help

- Automate workflows
- Enabling multiple services to communicate
- Work together efficiently

Kubernetes

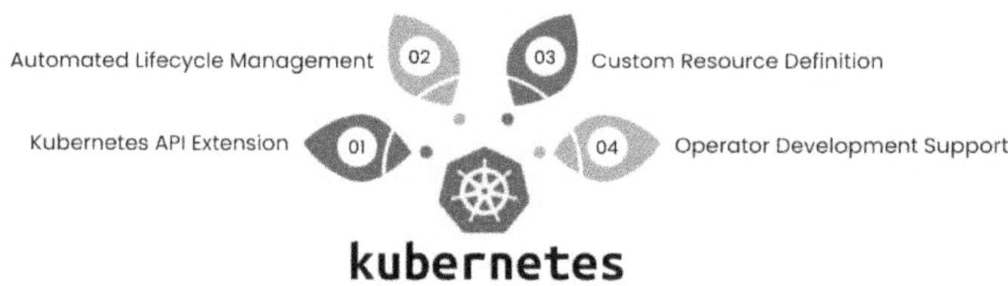

Figure 3.15: The image you sent shows the Kubernetes logo and highlights four key features of the Kubernetes platform for container orchestration: Automated Lifecycle Management, Custom Resource Definition, Kubernetes API Extension, and Operator Development Support.

Source: - *(Sinha, 2024)*

Kubernetes act as specialized controllers that expand the Kubernetes API to handle complex stateful applications, advancing Infrastructure as Code in DevOps. To effectively provide infrastructure resources, major cloud platforms such as AWS, Microsoft Azure, Google Cloud, and Oracle Cloud Infrastructure provide customized K8s operators.

Key Features:

- **Kubernetes API Extension:** Enhances the Kubernetes API to manage specific application needs, improving operational flexibility.

- **Automated Lifecycle Management:** Simplifies the management of complex stateful applications, ensuring seamless operation within Kubernetes clusters.

- **Custom Resource Definition:** Empowers teams to create and manage custom resources in Kubernetes, boosting control over infrastructure setups.

- **Operator Development Support:** Facilities building new operators, promoting innovation in infrastructure automation.

Provisioning Tools

Provisioning tools focus on the initial creation and configuration of infrastructure components like servers, networks, and storage. They automate the setup process, ensuring a faster and more reliable infrastructure deployment.

Terraform

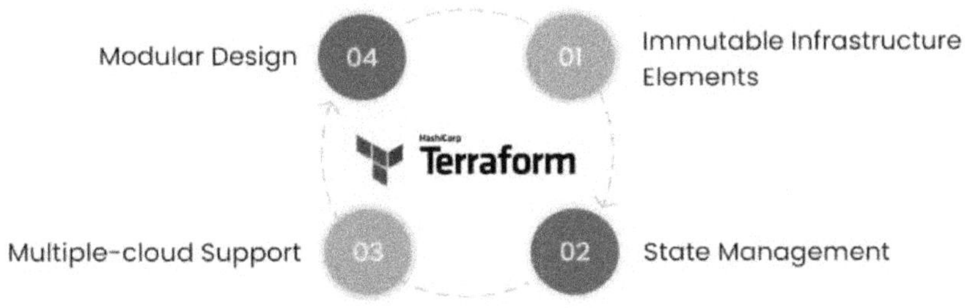

Figure 3.16: The image you sent shows the Terraform logo and highlights four key features of the Terraform platform for Infrastructure as Code (IaC): Immutable Infrastructure Elements, State Management, Multiple-Cloud Support, and Modular Design.

Source: - *(Sinha, 2024)*

Terraform stands as a leading tool in Infrastructure as Code in DevOps, enabling seamless infrastructure management across multiple cloud platforms using a straightforward declarative language.

Key Features:

- **Immutable Infrastructure Elements:** Ensures consistent configurations, reducing the risk of drift in infrastructure settings.

- **State Management:** Tracks infrastructure state to maintain stability and manage resource dependencies efficiently.

- **Multiple-cloud Support:** Offers compatibility with various cloud providers, enabling diverse infrastructure management.

- **Modular Design:** Promotes reusability and efficiency by allowing the creation of modular code components.

AWS CloudFormation

AWS CloudFormation offers a unified approach to defining and provisioning infrastructure within AWS cloud services, leveraging Infrastructure as Code in DevOps to streamline cloud resource management. This service, tailored exclusively for AWS, ensures a smooth and automated deployment experience, keeping pace with AWS's latest features and services.

Key Features:

- **AWS Integration:** Seamlessly aligns with AWS services, delivering optimizing support for a wide range of resources.

- **Declarative Templates:** Utilizes JSON or YAML formats to simplify infrastructure definitions.

- **Change Management:** Offers preview capabilities to assess and manage changes before applying them to your infrastructure.

- **Stack Organization:** Groups resources into manageable stacks, enhancing clarity and ease of resource handling.

- **Comprehensive Resource Support:** Extensively support AWS resources, ensuring compatibility with the latest offerings.

These tools play a critical role in Infrastructure as Code in DevOps, enhancing automation, efficiency, and consistency in managing IT infrastructure. They empower teams to build, deploy, and scale environments effortlessly, driving innovation and accelerating development cycles.

Best Practices for Implementation of IaC in DevOps

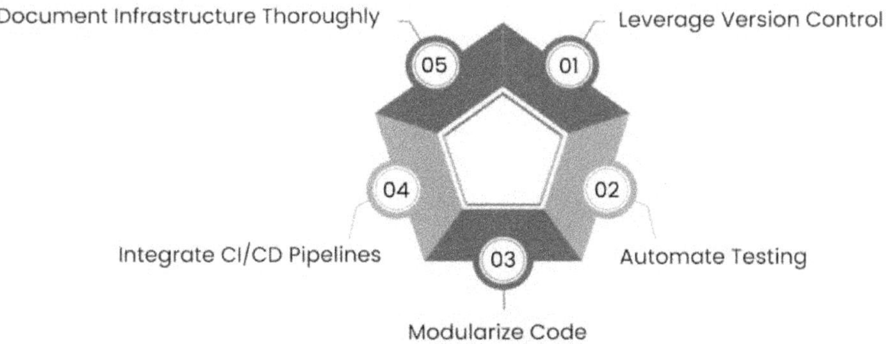

Figure 3.17: The figure illustrates best practices for implementing Infrastructure as Code (IaC). These practices include leveraging version control, automating testing, modularizing code, integrating with CI/CD pipelines, and thoroughly documenting the infrastructure.

Source: *- (Sinha, 2024)*

Let's delve deeper into the best practices for implementing Infrastructure as Code in DevOps, enhancing each point to maximize its effectiveness in a DevOps environment:

- **Leverage Version Control:** Integrating IaC configurations into version control systems like Git is crucial. This practice not only prevents rework by storing every infrastructure update but also provides a clear audit trail of changes. It allows multiple team members to collaborate seamlessly, managing access, tracking revisions, and rolling back to a previous state if needed. Version control acts as both a safeguard and a robust change management system

- **Automate Testing:** Automated testing is vital to maintaining the integrity of your infrastructure setup. Use tools like Terratest to test Terraform modules or AWS Config to check resource compliance against best practices. Automated tests catch issues early, preventing faulty configurations from disrupting deployments. By validating infrastructure before it's deployed, you ensure that changes won't lead to instability or errors in production environments.

- **Modularize Code:** Modularization means breaking IaC code into distinct, reusable components. Instead of having a monolithic script, create small, independent modules that focus on specific tasks. This modular approach simplifies updates since you only need to modify individual parts without affecting the entire codebase. It also promotes code reusability, making it easier to scale your infrastructure by reusing proven modules.

- **Integrate CI/CD Pipelines:** Integrating IaC into your CI/CD pipeline is essential for the continuous delivery of infrastructure changes. By embedding testing, validation, and deployment of IaC into the CI/CD process, you ensure that the infrastructure code is automatically validated before deployment. This integration keeps infrastructure and application code in sync, reducing discrepancies and accelerating delivery cycles.

- **Document Infrastructure Thoroughly:** Comprehensive documentation is the backbone of effective Infrastructure as Code in DevOps. Detailing configurations, architectural decisions, and setup procedures aids in knowledge sharing across the team. Proper documentation simplifies troubleshooting and onboarding, helping new team members understand the infrastructure's architecture and its dependencies quickly.

Implementing these best practices fortifies Infrastructure as Code in DevOps by enhancing collaboration, minimizing risks, and ensuring scalable and consistent deployments. This structured approach to managing IT infrastructure accelerates development cycles and drives innovation in dynamic cloud environments

3.5 AI-Powered Automation: Predictive Maintenance and Self-Healing Systems

AI-powered automation automates business operations by combining AI technology with other tools. This automation can take place via hardware, like robotic process automation (RPA) in the real world, or software, where AI systems evaluate data, learn from it, and make judgements.

AI automation processes and learns from vast volumes of data using AI techniques such as computer vision, natural language processing (NLP), and machine learning algorithms. An AI application may guide intelligent decision-making based on what it has learnt after processing that data and creating an AI model.

These abilities help drive innovation in many industries. In the healthcare area, for instance, AI is assisting in the development of new medications, while in the automotive sector, autonomous driving is advancing.

As AI becomes more efficient, it will better handle larger volumes of data and speed up the progress of different industries.

Benefits of AI automation in business processes

Tools that reduce and streamline work can make running a business easier. Businesses may minimise workloads and serve customers with AI automation.

Advantages of AI automation include:

- Increased productivity. Running a business without tools takes time. AI can handle tedious jobs like data entry and analysis so staff can focus on more important duties.

- Improved client experience. Customer service requires timely offers, which can be difficult if organisations must analyse enormous amounts of data. AI automation can analyse enormous amounts of client data to improve customer experience.

- Improved decision-making. AI automation can process commercial and industry data faster than humans. AI can assist firms in forecasting, predicting product trends, and learning about industry trends. This data can guide their choices.

Challenges of embracing AI automation

AI automation offers several benefits. Before adopting AI automation, consider its drawbacks.

- **Concerns about data privacy.** Third-party, cloud-based automation technologies can expose critical business and consumer data to privacy breaches. Users and companies should follow data privacy rules. Additionally, they should carefully evaluate third-party vendor security.

- **Labour market effects.** Routine employment may be replaced by AI automation. This may encourage affected workers to learn new skills. To adapt to AI, companies should train, reskill, and upskill their employees.

- **Challenges in implementation.** Implementing AI automation requires expertise and is complicated. Finding talent and recognising AI's limits may be difficult for companies.

What is Predictive Maintenance?

To anticipate equipment failure and plan remedial maintenance, predictive maintenance software uses data science and predictive analytics. The objective is to optimize equipment lifespan before it is compromised by scheduling maintenance at the most convenient and economical period. Data collection and storage, data transformation, condition monitoring, asset health assessment, prognostics, a decision support system, and a human interface layer are all common components of predictive maintenance systems.

Acoustic, corona detection, infrared, oil analysis, vibration analysis, sound level measurements, thermal imaging, and predictive maintenance all employ wireless sensor networks to collect and assess data in real time from machinery and activities. Equipment hazards are identified by predictive maintenance service providers with the use of these metrics and machine learning techniques such as regression and classification.

The following phases make up the process flow for predictive maintenance:

1. Determine equipment and failure mode to be monitored.

2. Establish frequency.

3. Monitor condition.

4. Issue report.

5. Is there an abnormality? If no, return to Step 3. If yes, continue to Step 6.

6. Create a work order.

7. Plan work date.

8. Ensure parts and labour are available.

9. Perform repair.

10. Close the work order and return to Step 3

The key components which support PM are IoT sensors to gather data, ML to analyze data, and cloud to store data and perform data analysis; in other words, PM involves tracking health of equipment in near real time, analyzing data to anticipate problems before they turn into catastrophic failures, and offering recommendations on when to perform maintenance.

Key points about these technologies:

- **IoT Sensors:** These devices retrieve critical performance parameters such as head and tail, temperature, pressure and current from mechanical systems, and offer real-time information of the status of the equipment.

- **Machine Learning:** The collected sensor data is fed to the ML algorithms, which then understand patterns, trends and outliers, and pre-emption of failures can be made.

- **Cloud Computing:** Repetitive data is produced by IoT sensors, and cloud platforms store and handle the massive amounts of data required for analysis and model building.

Other important aspects of predictive maintenance technologies:

- **Big Data Analytics:** Combining data from multiple sources to obtain information about the condition of equipment with a view of possible problems specific to them.

- **Edge Computing:** To improve reaction time and reduce network burden, the process should be moved closer to the data generation site.

- **Digital Twins:** To provide a digital version of an object in order to test various situations and fine-tune maintenance approaches on the physical object.

Benefits of using predictive maintenance:

Less frequency of breakdowns and extra equipment breakdowns, Efficient schedule management, Longer equipment life, saving of costs by avoiding unnecessary maintenance, and Better performance.

3.6 Chatbots and Virtual Assistants in Automated Cloud Services

Companies are using chatbots and virtual assistants more and more for customer service, marketing, and content creation. These computer applications communicate with people using voice or text interfaces.

AI technologies and chatbots are being used by modern businesses to enhance customer service.

Companies that use chatbots report significant gains in staff productivity and customer happiness. Eighty percent of users like the experience, and they assist personalise customer relationships. Chatbots, as opposed to human agents, are able to work around the clock, offering prompt support and promptly answering consumer questions. They free up human agents to work on more complicated problems by reducing their burden.

Benefits of AI Chatbots

Consumers today anticipate prompt and effective resolution to their questions and issues. Conventional customer service techniques, such as emails and phone conversations, are unable to satisfy this rising need.

Virtual assistants and chatbots improve consumer interaction, expedite procedures, and spur expansion. These are some of the main advantages of utilising customer service driven by AI:

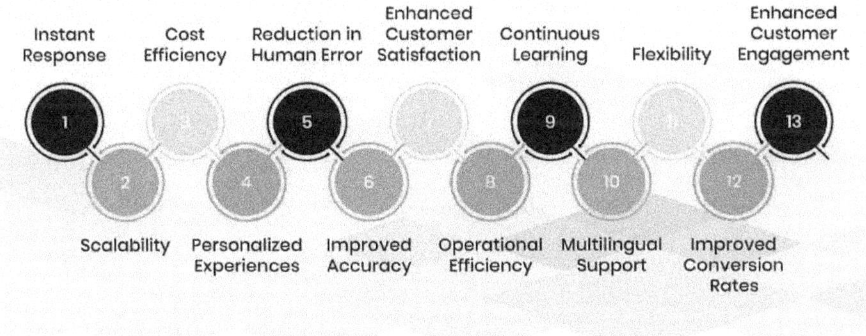

Source: - *(AnalytixLabs, 2022)*

- **Instant Response:** Real-time answers to consumer questions are provided by AI chatbots. To increase satisfaction, they expedite resolution and cut down on wait times.

- **Scalability:** Chatbots are essential AI customer service tools that can effectively respond to a wide range of questions and assist in managing surges in demand for customer care.

- **Cost Efficiency:** Chatbots lessen the effort for human agents and cut expenses by automating repetitive queries.

- **Personalized Experiences:** In order to provide personalized suggestions depending on the consumer journey, AI chatbots customize conversations using user data.

- **Reduction in Human Error:** Customer service is sped up by applying AI techniques to automate answers to frequently asked questions and FAQs.

- **Improved Accuracy:** Artificial intelligence (AI) systems for customer service minimize the possibility of providing clients with inaccurate information by generating answers from real knowledge sources.

- **Enhanced Customer Satisfaction**: Chatbots may answer several questions at once and are accessible around-the-clock to provide prompt, effective answers to consumer questions.

- **Operational Efficiency:** AI-powered customer support increases productivity by automating routine tasks, freeing up human agents' time to handle complex issues.

- **Continuous Learning:** AI chatbots gradually provide better service by learning from user interactions and refining their replies.

- **Multilingual Support:** AI customer support software enables chatbots to interact with clients in a variety of languages, which facilitates serving a wide range of clientele.

- **Flexibility:** Customers may communicate through their chosen channels with AI chatbots incorporated into a variety of communications platforms, including websites, social media, and mobile applications.

- **Improved Conversion Rates:** In addition to providing a round-the-clock sales funnel and serving as online sales assistants to improve the shopping experience, chatbots help move leads towards conversion.

- **Enhanced Customer Engagement:** Chatbots enhance consumer interaction by providing timely, individualized replies. It facilitates forging closer bonds with clients.

Challenges of Using AI Chatbots for Customer Support

While there are many advantages to using AI chatbots for customer service, there are also certain drawbacks that companies may encounter. A few such problems include:

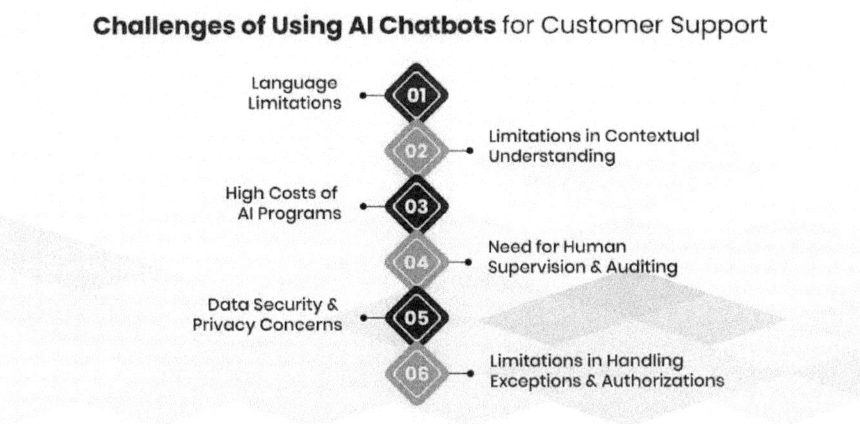

Source: - *(AnalytixLabs, 2022)*

- **Language Limitations:** Even with advances in natural language processing (NLP), chatbots may still struggle to comprehend and react appropriately to emotive and complicated language.

- **Limitations in Contextual Understanding:** AI chatbots may find it difficult to provide precise answers and contextualize data.

- **High Costs of AI Programs:** Sometimes the initial investment in chatbot setup, upkeep, and optimisation is outweighed by the time and resources required.

- **Need for Human Supervision and Auditing:** To guarantee the consistency, correctness, and moral use of client data, AI chatbots need constant oversight and monitoring.

- **Data Security and Privacy Concerns:** Concerns about data security, permission, and the appropriate use of sensitive information are brought up when integrating AI chatbots with CRM.

- **Limitations in Handling Exceptions and Authorizations:** AI chatbots could find it difficult to deal with exceptions, mistakes, or circumstances that call for judgement.

Applications of AI Chatbots in Customer Support

AI chatbots are flexible instruments that may be incorporated into a range of customer service tasks. Here's a closer look at how AI technologies are being used in customer service:

Source: - *(AnalytixLabs, 2022)*

1. Answering FAQs

Chatbots are excellent at answering often asked queries. They can use natural language processing (NLP) to comprehend the meaning of a customer's question, even if it is not expressed in the usual way. They enhance the customer experience by responding to recurring consumer enquiries with precision and speed.

A retail chatbot, for instance, may respond to frequently asked queries concerning product availability, return policy, and shop hours. It also responds to questions on product issues, delivery details, and order status. The chatbot may also swiftly direct users to the information they want, saving them from having to browse through several menus or web pages.

2. Scheduling Appointments

AI chatbots facilitate appointment scheduling by assisting users with booking, rescheduling, and cancellation without the need for human assistance. Chatbots may access real-time availability by linking with calendar systems, which enables clients to schedule appointments without requiring back-and-forth conversation with human agents.

Consumers only need to enter their desired time and date, and the chatbot will locate the closest time period that works for them.

For example, in the medical field, a chatbot may help people make appointments with doctors. It offers pre-visit advice to cut down on wait times and automatically sends reminders to decrease no-shows. Appropriate scheduling maximizes the time and resources of the healthcare professional while improving the patient experience.

3. Cross-Selling and Upselling

Chatbots work well for upselling and cross-selling. A customer's past purchases and browsing habits can be used by chatbots to find upgrades or related goods that the user would find interesting.

The chatbot can provide recommendations during encounters that are either additions to the customer's current choices or complementary goods to what they are presently buying.

For example, if a consumer buys a smartphone, the chatbot may suggest extras like wireless earbuds, screen protectors, or cases. These tailored suggestions not only increase sales but also improve the consumer experience by making it simpler for them to find goods they might need or desire.

Artificial intelligence chatbots boost client engagement and the likelihood of repeat business by providing timely and pertinent recommendations.

4. Order Processing

Chatbots help consumers place orders, verify the progress of their orders, and manage exchanges or returns. By interacting with backend systems, they may obtain data regarding delivery schedules, shipment information, and inventory levels. They lessen the burden for human agents while streamlining the ordering process and preventing cart abandonment. Customers may save time and effort by not having to call or email customer support for updates.

An e-commerce website's chatbot, for example, might assist clients with ordering by assisting them with product selection, colour and size options, and payment methods. Additionally, it may help with refunds, discuss expected arrival dates, and offer tracking information.

5. 24/7 Availability

Regardless of time zone, round-the-clock availability guarantees that client enquiries are addressed quickly. AI chatbots can work around the clock, unlike human workers who need breaks, sleep, and downtime. Day or night, they respond to consumer enquiries.

For multinational companies with clients in many locations, it is especially advantageous. 24/7 accessibility guarantees that clients may obtain help at any time, including after regular office hours.

A banking chatbot, for instance, may assist users with account balance checks, money transfers, and question answering day or night. It improves client satisfaction by doing away with the need to wait for a human representative during business hours.

6. Improving Response Times

Chatbots answer consumer questions promptly, cutting down on wait times and raising customer satisfaction levels. In the fast-paced market of today, it helps businesses keep ahead of the competition. Customers are more satisfied when response times are faster since they don't have to wait for help.

An AI chatbot, for instance, can be used by a financial services organisation to promptly respond to standard enquiries like branch locations, recent transactions, or account balances. Complex matters like loan enquiries or fraud detection can be handled by trained human agents.

7. Ensuring Consistency

Chatbots that provide a consistent quality of service and standardised responses to consumer enquiries are used in AI-powered customer care. In contrast to human operators who may have different levels of expertise and temperament, chatbots consistently provide customer service. It guarantees that all clients get information that is correct and consistent. When interacting with customers, consistency fosters dependability and trust.

A telecom chatbot, for instance, can reliably answer enquiries concerning technical assistance, billing concerns, and service plans. Additionally, chatbots may be updated often with the most recent data, guaranteeing that clients receive accurate and current assistance.

This reduces the possibility of misunderstandings and enhances the customer service process's overall effectiveness.

8. Routing Complex Queries

In order to ensure that consumers obtain specialised assistance when needed, AI systems for customer care are skilled at effectively forwarding complicated enquiries to the relevant human agents. Advanced natural language processing (NLP) skills enable chatbots to precisely understand and classify consumer enquiries.

The chatbot instantly recognised the need for human involvement when a query or problem exceeds its preprogrammed capabilities. It passes the question to a human agent who is most qualified to answer it with ease. The hybrid strategy guarantees that delicate issues are handled properly while optimizing resource allocation.

A financial services chatbot, for example, would respond to standard questions but forward those pertaining to loan applications or investing advice to a human adviser

Furthermore, by giving the human agent context and pertinent information acquired during the first conversation, the chatbot may expedite the resolution process and reduce the need for clients to repeat themselves.

9. Multitasking

AI-powered customer support tools able to multitask by processing orders, answering questions, making product suggestions, and setting up appointments. They are made to quickly transition between different duties, guaranteeing that every client gets help when they need it.

It decreases wait times and increases efficiency. Businesses may handle large amounts of enquiries with this capacity without sacrificing the quality of their services.

For example, at a product launch, a chatbot may accept pre-orders, answer questions about the new product's features, and give early adopters troubleshooting advice all at once.

Complex or high-priority problems that need a human touch can be handled by human agents. AI chatbots contribute to the development of a responsive and well-balanced customer care system that can readily adjust to changing demands by sharing the burden.

10. Personalization

AI and ML algorithms allow chatbots to retrieve information from prior conversations, browsing habits, and purchasing history to give a tailored experience. From greeting returning customers by name and retaining their preferences to delivering personalized recommendations, customisation creates better customer connections.

Additionally, chatbots may provide highly personalized marketing messages and advertising campaigns by utilising data about client preferences. Engagement and conversion are enhanced.

An e-commerce chatbot, for instance, can identify a repeat consumer by name and provide product recommendations based on their browsing or past purchases. If a consumer regularly purchases running shoes, the chatbot may suggest the newest styles or alert them to impending sporting gear specials.

Product suggestions are only one aspect of personalization. By storing client preferences for language, channels, and even preferred conversation times, AI chatbots may also customize the support experience.

If a client often uses a mobile app to communicate with the chatbot in the evening, for example, the chatbot may priorities providing proactive support messages or notifications at that time. Customers will receive timely and pertinent information thanks to this degree of personalization, which improves their entire experience and happiness.

3.7 Chapter Summary

In this chapter, the author examines the nature of the change brought by browsers and bots in automated cloud services, particularly as a means of support, advertising, and information delivery for customers. It focuses on the major advantages of using AI to implement chatbots as instruments for interaction with customers, including faster response rates, the possibility to raise the handling of many customers and issues at once, reduced costs per customer, and highly personalize communication patterns that can lead to greater satisfaction for customers and efficiency for businesses. Chatbots are online all the time, which can respond to more than one user's questions and will not tire or grow incompetent as a human operator may. The chapter also brings out the risks that are associated with the adoption of chatbots such as; Limited language capability, lack of contextual understanding, data security issues and the expenses that comes with integration of chatbots. Real-life application scenarios, like the management of questions and answers, appointment setting and calendar, and order taking and management are all essential ways AI chatbots enthral users and bring value to the business. Altogether, chatbots and virtual assistants provide a critical function in the maintenance and development of customer service functions, as well as the improvement of the general efficiency and customer experience.

Multiple Choice Questions (MCQs)

1. **What is the primary purpose of automation in cloud computing?**

    a. Increase manual control over resources

    b. Reduce human intervention and streamline operations

    c. Eliminate the need for software development

    d. Replace cloud security measures

2. **Which of the following best describes DevOps?**

    a.   A programming language for cloud automation

    b.   A framework for managing virtual assistants

    c.   A cultural shift combining development and operations for continuous delivery

    d.   A cloud storage system

3. **What does CI/CD stand for in cloud automation?**

    a.   Cloud Integration/Cloud Deployment

    b.   Continuous Integration/Continuous Deployment

    c.   Centralized Infrastructure/Cloud Development

    d.   Containerized Installation/Code Deployment

4. **How does Infrastructure as Code (IaC) simplify cloud management?**

    a.   By reducing the need for automation

    b.   By using code to provision and manage infrastructure automatically

    c.   By manually managing cloud resources

    d.   By eliminating resource scaling

5. **Which of these technologies is commonly used for predictive maintenance in AI-powered automation?**

    a.   Machine Learning models

    b.   VPN Services

    c.   Cloud Storage Providers

    d.   IoT Devices

6. **What is a key benefit of using self-healing systems in cloud environments?**

   a.   Increased need for manual intervention

   b.   Automatic error detection and correction

   c.   Reduced data storage

   d.   Limiting automation capabilities

7. **How do chatbots contribute to cloud automation?**

   a.   By managing cloud infrastructure

   b.   By enhancing customer support through automated responses

   c.   By reducing cloud security

   d.   By manually processing user data

8. **What role does version control play in DevOps automation?**

   a.   It secures cloud storage

   b.   It tracks code changes and ensures collaboration

   c.   It replaces infrastructure automation

   d.   It eliminates the need for cloud services

9. **Which tool is commonly associated with Infrastructure as Code (IaC)?**

   a.   Tableau

   b.   Jenkins

   c.   Terraform

   d.   Docker

10. **What is the key focus of AI-powered automation in cloud contexts?**

   a. Manual data processing

   b. Human-controlled infrastructure management

   c. Automatic optimization and performance improvements

   d. Eliminating cloud services

Answer

1	2	3	4	5	6	7	8	9	10
b	c	b	b	a	b	b	b	c	c

Chapter 04

CLOUD SECURITY AND COMPLIANCE IN AI-DRIVEN ENVIRONMENTS

4.1 Chapter Overview

Chapter 4 focuses on the important elements of cloud security and compliance in the context of AI integration to understand the importance of the proper measures that must be taken to protect data and meet the essential requirements. It describes the main security concerns associated with AI technologies, including data and model exposures, threats linked to distributed cloud environments. Some of the basic topics for discussion include; Encryption and decryption, Approaches to access control, Data anonymization techniques, and Decentralized AI model deployment. The chapter also discusses such practices and acts such as GDPR, HIPAA as well as the ISO to show how compliance in the use of data can be achieved when using cloud-based AI systems. Finally, this chapter explains how to manage the tension between innovation and protection, and how to maintain the trustworthiness and compliance of AI cloud solutions.

4.2 Cloud Security Architecture and Threat Mitigation

What is cloud security architecture?

The structure and technology used to safeguard data and systems in the cloud are known as cloud security architecture. Your information is safe and secure in a variety of cloud settings thanks to this framework's collection of security mechanisms that assist stop illegal access and data breaches.

Why is cloud security architecture important?

The security architecture of cloud computing is important for a number of reasons. Businesses use cloud services to store data in cloud containers or to host workloads and apps. eCommerce sites are powered by PaaS systems, while some businesses have moved to IaaS systems for their whole IT infrastructure.

External attackers must be prevented from accessing data that is either in transit or at rest on cloud servers. Otherwise, hackers might get access to private data flows and take sensitive data out. DDoS assaults may seriously impair business operations.

Cloud assets must be protected against insider threats and unsafe employee conduct by security teams. Cloud assets can be immediately compromised by weak passwords and a lack of understanding about phishing. To safeguard cloud data, businesses must consider local data compliance challenges. Additionally, they must monitor device security to make sure all linked devices are secure.

The architecture of cloud security is the responsibility of users. They are unable to work app-by-app or delegate security to cloud partners. Rather, best practices for cloud security advise developing enterprise-wide plans to address all cloud security threats.

To put it another way, businesses need to design an architecture that safeguards important resources. A catastrophic vulnerability in any enterprise security posture is a poorly designed cloud security architecture. However, how does cloud security architecture actually work? Let's investigate how to create that framework and safeguard cloud assets.

Components of A Cloud Security Architecture

The architecture of cloud security is not a one-size-fits-all solution. Although there are differences in cloud implementations, Software-as-a-Service, Platform-as-a-Service, and Infrastructure-as-a-Service are all often utilised. The apps being used, how businesses employ cloud apps, and who needs access all affect security measures. Depending on the situation, different instruments are needed. Nonetheless, practically every cloud security plan includes a few essential technologies:

- **Encryption:** Data travelling between network users and cloud resources is jumbled by encryption. Attackers are unable to read messages or obtain important information while data encryption is in place. From user devices to cloud gateways, all data is secured from beginning to end. Data on your cloud provider's services should be encrypted while it's at rest. Encryption for virtual private networks (VPNs) may also be applicable to network infrastructure. VPNs protect users' identity from prying eyes. By adding an additional layer of encryption, they make data theft even more difficult.

- **Access management:** Cloud resource access is managed using Identity and Access Management (IAM) solutions. Client-side access control serves as a bridge connecting cloud apps and corporate networks. Each user's access to various cloud resources is determined by the rights that IT managers provide to them. Multi-factor authentication (MFA) may also be included into IAM to improve access security.

- **Visibility tools:** Cloud-connected devices need to be always visible. Cloud Access Security Brokers (CASBs) send out notifications regarding anomalous activity and offer real-time information. In order to identify potential risks, security professionals can keep an eye on user behaviour and data flows. Additionally, they have the ability to record user activities for audit and compliance purposes. Network managers may now enhance their cloud security architecture thanks to this.

- **Automation:** Cloud security management does not require labor-intensive monitoring. Security teams can automate monitoring systems to run in the background. Additionally, companies may automate security policy maintenance and software updates, freeing up time for cybersecurity analysis.

- **Network segmentation:** Cloud apps are separated from other network resources by segmentation. A zone of trust is established around each cloud gateway through the segmentation of cloud assets. Unauthorized actors are not allowed in certain areas; only verified users are allowed. If hostile actors infiltrate the network through vulnerable endpoints, segmentation stops them from accessing cloud resources. In addition to security features like IAM and activity tracking, it offers Zero Trust Network Access (ZTNA) as a basis.

- **Flexibility and scaling:** Strategies for cloud security must be flexible enough to grow and adjust when network perimeters shift. When introducing new cloud services to the network, security teams need to be able to expand security protection. Cloud resources should be compatible with user onboarding and offboarding. When starting new positions, employees should have access to essential resources, and automated systems should delete orphaned accounts right away.

- **Resilience:** The design of cloud security must incorporate strategies to increase the resilience of cloud systems. Cloud installations require firmware that is resilient to cyberattacks. Critical cloud resources must be quickly restored by businesses in the event of an attack. To avoid malware infection when cloud systems are starting up, security teams should additionally strengthen the boot procedure.

Software for cloud security posture management (CSPM) can aid in compliance. Security settings are compared to compliance best practices using CSPM tools. They make it easier to accomplish regulatory targets by highlighting any necessary actions.

4.2.1 Cloud Security and Shared Responsibility Model

Cloud security design may be comprehended by breaking the problem down into SaaS, PaaS, and IaaS methods. In all three situations, users and cloud providers follow the shared responsibility approach.

Source: - *(Nordlayer, 2023)*

The cloud service provider is in charge of protecting apps, including the code, hardware, and infrastructure needed to provide client services, according to the shared responsibility model. However, this implies that users are in charge of the information they save on cloud servers. Not every cloud user understands what this implies.

- **Infrastructure as a Service (IaaS):** Customers purchase cloud infrastructure from a third-party cloud provider in IaaS solutions such as Microsoft Entra ID. Businesses are able to integrate their operating systems and software into this architecture. This is an excellent method for developing a personalized cloud deployment. Customers are in charge of protecting the apps they install on cloud infrastructure under this approach. The security function of cloud providers is rather little.

- **Platform as a Service (PaaS):** Customers purchase off-the-shelf platforms that operate on pre-built cloud infrastructure in PaaS solutions. Typical examples are SAP Cloud, Heroku, and Amazon Web Services. Customers using PaaS platforms are required to safeguard their apps under this agreement. They are in charge of controlling access and configuring the app. Compared to IaaS partners, cloud providers need to develop safe systems and take a more proactive approach to security.

- **Software as a Service (SaaS):** On a case-by-case basis, SaaS entails buying access to cloud-based apps. Salesforce, Zoom, Dropbox, and Mailchimp are all excellent examples. Businesses that use SaaS applications are in charge of controlling user access. Security personnel must verify users and weed out bad actors. App security is the responsibility of cloud partners.

4.2.2 Principles of Cloud Security Architecture

It is a good idea to base your strategy on fundamental ideas when developing a cloud security architecture. Three principles— confidentiality, availability, and integrity—apply in the cloud setting. Your cloud security strategy should incorporate all three of these ideas.

- **Confidentiality:** Cloud-based data should be hidden from other parties. Only authorised users should be allowed to access data. Businesses ought to restrict access based on trust. Confidential information should only be accessible or editable by those who possess the necessary authorization. Security teams should examine each user's rights on a frequent basis.

- **Availability:** There should always be access to cloud-hosted apps and customer support. Security teams must eliminate availability threats.

DoS assaults pose a serious risk in this field, but security personnel must counteract any threats that have the ability to stop services.

- **Integrity:** Businesses want assurance that data and apps work as intended. Additionally, they want confirmation that apps haven't been altered without authorization. Without authentication, users or hackers shouldn't be able to alter data and application code. Tools should monitor any modifications and identify problems with data integrity as soon as they appear.

4.2.3 Cloud Security Architecture Threats

Building a cloud security architecture is difficult without an understanding of the main risks to cloud security. For cloud users to manage the most serious cyberthreats, risk strategies and mitigation techniques are essential. Cloud users should take into account the following common risks:

- **Denial-of-Service (DoS) attacks:** Cybercriminals send massive volumes of bandwidth to cloud apps in denial of service assaults. System overload eventually causes network degradation and program termination. Security technologies can keep an eye out for indications of DoS assaults before they become serious and filter possible DoS traffic at the network edge. Cloud partners could also have measures in place to reroute traffic in the event of a DoS attack.

- **Insider attacks:** Many cloud attacks are carried out by insiders of the firm. Employees may inadvertently give login information to third parties or abuse it. Employee accounts are frequently compromised by phishing and weak password security. Additionally, cloud companies could harbour insiders who have access to customer information. Additionally, unprotected cloud assets might be subject to unofficial or governmental surveillance.

- **Poorly secured cloud hardware:** Cloud partners could make use of storage infrastructure that is spread out throughout the globe. Additionally, they could outsource out hardware management to other businesses. Network edge security flaws may result from this.

Physical hardware, for example, might not be adequately secured by third parties. For attackers, this makes things much easy.

- **Shadow IT:** Cloud settings could not be entirely beyond the control of security professionals. For instance, IT personnel may not be aware of changes made to app configurations. This issue is frequently linked to DevOps teams. Because shadow IT might result in security flaws, stack validation procedures are crucial.

- **Human error:** Simple errors might potentially have an impact on cloud security. For instance, users frequently configure S3 Storage Buckets incorrectly, giving hackers a point of access. Security control panels may be left inadequately guarded by businesses. It is possible for user permissions to become too biassed towards user access rather than security.

Create A Rock-Solid Cloud Security Architecture

An essential component of contemporary corporate management is the security architecture of cloud computing.

Large volumes of data pass via cloud apps and containers every day. SaaS apps facilitate private doctor-patient consultations and important work meetings. Payroll management is handled by HR apps, whereas sales and recruiting apps hold private financial information. These cloud services all need to be protected from outside attacks.

Confidentiality is safeguarded by a strong cloud security architecture. It makes services available to stakeholders and guarantees data integrity. PaaS, IaaS, and SaaS systems must all be well-suited for security measures. They must also take into consideration their shared accountability with cloud partners.

Fortunately, cloud security configurations are aided by a variety of solutions. Backups, disaster recovery tools, segmentation, activity tracking, VPNs, CASBs, encryption, and access management consoles all have functions. Develop a plan that works for you.

4.3 Data Privacy and Compliance Regulations

Maintaining a cloud-based business requires an awareness of the complexities of cloud compliance, which is becoming increasingly important as businesses continue to move their operations to the cloud. Cloud compliance is more than simply following rules; it's a thorough strategy for handling operational integrity, data security, and privacy. As cybersecurity threats targeting cloud-based infrastructures increase, maintaining compliance is essential to safeguarding a business and its clients.

According to our 2023 Currents study report, which polled tech founders, executives, and workers, 37% of them said they want to spend more on cybersecurity in the next fiscal year. 34% of that group said that the emergence of generative AI presents new cybersecurity risks, and 0% reported having recently encountered a security event. By protecting sensitive data and maintaining the integrity of cloud systems, cloud compliance is an essential defence against more complex data breaches and cyberattacks. The key components of cloud compliance will be covered in this article, along with how it affects cybersecurity and why it's crucial for cloud-based businesses.

What is cloud compliance?

Adhering to a set of regulations and standards necessary for cloud computing services is known as cloud compliance. These rules, which are usually set by government agencies, trade groups, or corporate policies, ensure that data handled and stored in the cloud is safe and used responsibly. Cloud compliance maintains confidence between cloud service providers and their clients while safeguarding sensitive information and private data. Compliance is crucial for operational security and customer trust, especially in light of the rise in data breaches.

Components of Cloud Compliance

Making ensuring that cloud usage complies with different legal, regulatory, and corporate norms requires a complex strategy known as cloud compliance. Understanding and putting into practice the essential elements that regulate data management and security in the cloud environment are necessary to achieve cloud compliance. Any company using the cloud must have these elements in order to comply with regulations and preserve data integrity.

- **Standards:** A variety of best practices and benchmarks that specify how cloud services should be utilised and maintained are included in standards for cloud compliance. These include information security management frameworks such as ISO/IEC 27001, which offer a methodical way to handle confidential enterprise data. Following these guidelines guarantees the security of cloud infrastructure and the effectiveness of cloud operations for companies. Adherence to these guidelines improves security and increases client confidence in cloud service providers. To satisfy cloud compliance requirements, cloud service providers must make sure their offerings are in line with these standards.

- **Laws and regulations:** A key component of cloud compliance is laws and regulations, which specify the legal requirements for companies that use cloud services. The General Data Protection Regulation (GDPR), which enforces stringent guidelines on data protection and privacy for individuals inside the European Union, is a notable illustration. Businesses must comprehend and abide by these rules in order to avoid fines and keep the trust of their clients. This means making sure their cloud provider abides by the applicable rules, particularly when it comes to managing private or sensitive data. Because rules change over time, navigating these laws is an ongoing effort.

- **Governance:** The policies and processes that companies set up to manage their cloud usage are referred to as governance in cloud compliance. Cloud services are employed in a way that satisfies company goals and regulatory requirements thanks to effective governance. This entails establishing precise guidelines for risk management, access control, and data security in the cloud environment. Robust governance frameworks help cloud service providers achieve cloud compliance standards on a regular basis. Additionally, it entails routinely reviewing and modifying these rules to conform to changing business requirements and compliance environments.

- **Audits:** An essential part of cloud compliance is audits, which give companies a way to confirm that their cloud services and infrastructure adhere to relevant laws and standards. These audits frequently entail a careful analysis of the cloud provider's data processing, storage, and security procedures. Frequent audits assist companies in finding areas for cloud use enhancement and compliance gaps. They are also an essential instrument for proving compliance to stakeholders and authorities. Maintaining continuous cloud compliance requires selecting a cloud provider that supports and allows frequent compliance audits.

Why is cloud compliance important?

Cloud compliance is essential for any cloud-based organisation. It guarantees that cloud services adhere to ethical and legal norms and are utilised responsibly. For companies, maintaining cloud compliance means protecting their operations, brand, and clientele. It goes beyond simply adhering to regulations.

- **Data protection and privacy:** Protecting sensitive data from breaches and unwanted access requires effective cloud compliance procedures. Businesses may lower the risk of data leaks and the related legal and reputational consequences by following compliance rules, which guarantee that sensitive and personal information is handled securely.

- **Legal and regulatory adherence:** Businesses that operate in the cloud must abide by rules and regulations like the GDPR. Legal penalties and fines, which may be severe and detrimental to a company's financial stability and reputation, are avoided by this commitment.

- **Business continuity and risk management:** Planning for business continuity and risk management both heavily rely on cloud compliance. It assists in recognising and reducing the risks connected to cloud use, making sure that compliance issues or data security breaches don't interfere with corporate operations.

- **Competitive advantage:** Compliance may be a differentiation in a market where many companies use cloud services. Businesses that exhibit strong compliance procedures might obtain a competitive advantage by drawing in clients and partners that respect ethical behaviour and data protection.

Best practices for cloud compliance

Cloud security and data management must be approached strategically in order to achieve cloud compliance, which is a continuous effort. Adopting best practices is essential for companies using cloud services to make sure they comply with legal obligations and safeguard sensitive data.

1. **Understand the shared responsibility model**

   Compliance in cloud computing is a joint duty between the customer and the cloud provider. In general, the client is in charge of protecting the data they store in the cloud, even while the cloud provider guarantees the security of the cloud infrastructure. Businesses must completely comprehend their role in upholding compliance under this shared responsibility approach, especially in areas like data encryption and access control. Potential weaknesses in security and compliance procedures can be found with the aid of a thorough grasp of this common model.

2. **Assess your compliance requirements**

   Every company has to carefully evaluate its unique compliance needs in light of its activities and data. This entails being aware of the pertinent laws and rules, such as GDPR for businesses that handle the data of EU residents or HIPAA for data pertaining to healthcare in the United States. It is essential to update these evaluations on a regular basis since company operations and the regulatory environment might change, requiring adjustments to compliance strategy.

3. Know the security risks specific to your business

Effective cloud compliance requires recognising and resolving the particular security threats relevant to your industry. The Payment Card Industry Data Security Standard (PCI DSS), for instance, must be strictly adhered to by companies in the financial technology industry that handle credit card transactions. To secure cardholder data, this entails putting strict data security measures in place, including as encryption, access restriction, and frequent security audits. Maintaining compliance and protecting sensitive financial data need an understanding of and attention to these sector-specific risks, such as possible weaknesses in transaction processing or data storage.

4. Protect your data through encryption

One essential procedure for protecting data in a cloud setting is encryption. Businesses may drastically lower the risk of illegal access and data breaches that necessitate a security incident response by encrypting data both in transit and at rest. It is particularly important to use strong encryption techniques when working with private or sensitive data. Furthermore, it's critical to handle encryption keys safely, making sure that only authorised individuals can access them and that they are shielded from outside dangers.

5. Understand your service level agreement (SLA)

One important document that describes the conditions of service, including performance criteria, uptime, and data management rules, is the service level agreement (SLA) that you have with your cloud service provider. To make sure that their SLA satisfies their cloud compliance standards, businesses must carefully examine and comprehend it. In addition to helping to establish clear expectations and obligations, this knowledge makes it obvious what redress and mitigation measures are available in the case of a service outage or data breach.

What Is Data Privacy Compliance?

The collection of procedures, guidelines, and rules used to shield a database from unwanted access, alteration, or destruction is known as database security. Database security rules are intended to guarantee the availability and integrity of records kept in a database system and to stop sensitive data from being exposed. Since unauthorized data modification and data breaches may cause serious operational and financial harm, many businesses are now placing a high priority on database and data storage security.

Database Security in Public Clouds

Database services can be controlled or left unmanaged by public cloud providers. Customers would be responsible for maintaining unmanaged or semi-managed databases on a virtual machine, while cloud providers are in charge of deploying security patches, upgrading software, and guaranteeing high availability for managed databases. (The cloud provider would still be in charge of managing the actual infrastructure.)

The shared responsibility approach governs cloud database security in both situations. Cloud providers are always in charge of protecting the underlying infrastructure, which includes the computer, storage, and networking resources. They would also be in charge of patching, upgrading, and keeping an eye out for any security threats with managed database services.

Organisations, on the other hand, are in charge of protecting the information kept in the databases, putting in place the required access restrictions, and adhering to all applicable laws. This entails setting up database access rights, encrypting private information, keeping an eye on questionable activity, and educating staff members on security best practices.

Organisations would frequently need to adopt agentless security measures since they frequently do not have access to cloud database servers. With no software installed on the actual database, these technologies monitor the database remotely via APIs and data extracts. Another benefit of agentless security products will be their reduced influence on cloud resource use and performance. Real-time threat detection, vulnerability scanning, and compliance monitoring are all possible with efficient agentless security solutions, protecting databases without requiring extra resource overheads.

Elements of Database Security

- **Authentication and Identity Management:** Implementing robust authentication techniques like multifactor authentication (MFA), single sign-on (SSO), and appropriate role-based access control (RBAC) can guarantee that only authorised users may access the database. This makes it possible to have fine-grained control over user rights and database access to private information.

- **Data Encryption:** encrypting data using industry-standard methods to protect it both in transit and at rest. This helps guarantee that the data will remain unreadable without the necessary decryption keys, even in the event of unauthorized access.

- **Data Masking and Redaction:** protecting sensitive information from users without the required authorization by using redaction or data masking techniques. Sensitive information can be secured even when accessible by authorised persons with restricted access by using techniques like anonymization, pseudonymization, or data obfuscation.

Intrusion Detection and Prevention

Installing intrusion prevention and detection systems (IDS and IPS) will keep an eye on the database and guard against malicious activity. These tools can assist in identifying and stopping attacks like cross-site scripting, SQL injection, and other efforts to take advantage of flaws in the database system.

Backup and disaster recovery: To reduce the effect of data loss or corruption brought on by hardware malfunctions, software bugs, or malevolent acts, regularly backup your data and develop a strong disaster recovery plan. This enables the company to promptly return the database to a safe and functional state after an event.

- **Patch Management:** Maintaining database software patches and upgrades helps guard against known security holes and vulnerabilities. By ensuring that updates are implemented consistently and on schedule, a patch management policy lowers the possibility of exploitation.

- **Access Control:** Granular access control measures, such the least privilege principle, are put into place to make sure users are only given the rights essential for their roles. This limits what each user can do within the database, reducing the possibility of illegal access or data modification. (See: What is data access governance?)

- **Network Security:** protecting the database's network architecture through the use of firewalls, virtual private networks (VPNs), and other security tools. This helps keep the database safe from unwanted access by protecting the lines of communication between the database and the other systems in the company.

Database Security: 8 Best Practices

Organisations may prevent data exfiltration and unwanted access to their databases by putting basic database security best practices into operation. Among these great practices are:

1. Regularly review and revise database security policies and procedures.

2. To find and fix any weaknesses in the security posture, keep an eye on database activity and provide security reports.

3. To stop unwanted access, use MFA, RBAC, and strong passwords.

4. To prevent unwanted access or alteration, encrypt data both in transit and at rest.

5. To guarantee database software is current and safe from known vulnerabilities, apply patches and upgrades as soon as possible.

6. Create a disaster recovery plan and make frequent backups.

7. To find any weaknesses in the database security posture, perform vulnerability scans and risk assessments.

8. Train employees about database security's significance and their part in preserving it.

4.4 Secure AI Workflows in Cloud Environments

Artificial Intelligence (AI) has become an integral part of modern business strategies, driving automation, innovation, and insights across various industries. As organizations increasingly deploy AI workloads in both cloud and on-premise environments, ensuring their security has emerged as a critical challenge. These workloads involve handling vast amounts of sensitive data, building complex models, and performing real-time analytics — all of which make them attractive targets for cyber threats.

Securing AI workloads is distinct from traditional security practices due to the unique nature of AI systems. Issues such as data poisoning, model theft, and adversarial attacks pose significant risks that demand specialized strategies.

This article will explain the security challenges associated with AI workloads and offer insights into best practices for protecting these critical assets, both in cloud and on-prem environments.

Additionally, we'll highlight AccuKnox's role in AI security and Go Cloud's services to help businesses overcome the complexities of AI workloads.

Understanding AI Workload Security in Cloud and On-Prem

Complex programs that carry out tasks like image recognition, natural language processing (NLP), and predictive analytics are known as AI workloads. Unlike traditional workloads, AI systems rely heavily on data — both structured and unstructured — requiring constant updates and access to a large pool of computing resources.

This makes them highly sensitive to disruptions, data breaches, and tampering, and that necessitates security measures.

- Cloud-based AI Workloads are often deployed in multi-tenant environments, utilizing the elasticity of cloud infrastructure to scale as needed. This makes them agile but also introduces risks related to shared resources, data leaks, and compliance issues.

- On-Prem AI Workloads, on the other hand, offer more control and are typically hosted within a company's private data centers. While this setup minimizes the risks associated with multi-tenancy, it can be more vulnerable to physical security threats, internal breaches, and challenges in integrating the latest security patches.

AccuKnox's Role

For these scenarios, Accuknox's Model Knox Solutions enables secure deployment and management of AI models, including cloud-native security benefits with local infrastructure privacy.

Model Knox, an AI/LLM security feature by AccuKnox, is designed to protect Generative AI (GenAI) and Language Learning Models (LLMs) against emerging vulnerabilities like data leakage, prompt injections, and unauthorized access. It applies a Zero-Trust model to secure these AI environments, embedding security measures that ensure LLMs handle data securely and predictably across diverse applications, particularly in GenAI workflows.

Challenges of securing AI workflows in cloud environments include a need to incorporate strategies that will ensure that data is protected, privacy as well as the quality of the model is maintained throughout the AI life cycle. Since AI systems deal with a large amount of data they also deal with sensitive data and therefore maintaining strong security measures is crucial to avoid data leaks, unauthorized access, and tampering of the models. Cloud environments which provide scalability and flexibility for AI workloads add more risks because of data leakage, can misconfiguration occur and compliance issues. In order to manage these risks organizations require a well-established security plan that addresses data encryption, and access control, monitoring and inspection.

One of the essential elements of AI workflow protection is data encryption not only while stored but also during its transfer. Current CSPs incorporate encryption features and KMVs for encrypting the data during the training and inference of AI models. Another best practice that can be put in place regards the protection of data and models is the use of role base access control (RBAC). Moreover, to enhance security, Multi-Factor Authentication (MFA) is really effective because the person needs to complete some multiple steps in order to access the restricted resources for the organization. SIEM systems can be used for monitoring access logs and in real-time to look for security threats.

Another important factor is the need to protect the AI model since it can be easily attacked, its intellectual property rights violated. To protect models from unwanted tweaks and attempts at reverse engineering, several measures are possible, such as versioning and storing of models, as well as the use of differential privacy techniques. In addition, the secure development pipelines with integrated CI/CD guarantee that all the AI components pass the security test before their deployment into the production environment. Adherence to norms for example ISO/IEC 27001 and SOC2 can also assist an organization to prove that their security measures are effective in cloud settings.

Finally, cloud AI operations are subject to constant risk evaluation and management of secure AI workflows. Following are the proactive Security measures that should be included in the development Security framework: Here again there is need for continuous and close cooperation between security specialists and data analysts to balance performance gains and security needs. Moreover, compliance with the recent changes in regulation policies such as GDPR and HIPAA is a great way to make AI activities correspond to the data protection requirements. It is, therefore, possible to adopt a comprehensive security perspective and use the potential of cloud platforms for AI solutions without security and compliance risks.

4.5 Identity and Access Management (IAM) Best Practices

The security requirements of digital enterprises have grown dramatically since the advent of cloud computing. Systems are at danger because outdated access control techniques are no longer effective. The move to contemporary cloud-based Identity and Access Management (IAM) architecture is being driven by the difficulty and expense of maintaining historical, on-premise IAM. However, how should it be done? To make it simple and prevent data breaches, let's go over IAM best practices.

Identity and Access Management overview

The technologies we employ to control access to network resources both inside the network edge and at the perimeter are referred to as identity and access management.

Identity, which comprises authentication and controlling network entry points, and access, which entails establishing particular user rights and regulating user mobility within network resources, are the two main components that modern IAM systems oversee.

IAM seeks to protect the rest of the network from unwanted access while allowing users to access the resources they need. It frequently works in tandem with Zero Trust Network Access rules to guarantee adherence to sector-specific security standards.

Well-executed modern IAM systems can grow to meet new demands, adjust to changing network perimeters, and minimize the threat surface accessible to cybercriminals. Mistakes, however, can result in inefficiencies, additional expenses, and—more concerning—wide-ranging security flaws.

Fortunately, cloud and hybrid-based IAM may be implemented in a variety of ways that combine security and simplicity.

9 Identity and Access Management best practices

Clear objectives, astute privilege mapping, and a Zero-Trust stance are all necessary for effective IAM. A strong system relies on automation, frequent audits, and compliance.

Nine essential steps have been identified to guide organizations in implementing secure and streamlined access across their operations.

IAM Best Practices

Source: *- (NordLayer, 2022)*

1. Understand project goals

Visualizing the endpoint of your IAM project is crucial first. There are several justifications for putting Identity and Access Management into practice.

Requests to reset passwords may overload customer support staff. Concerns may arise over current and possible internal sabotage or phishing threats, and security audits may have shown security flaws such as granting users inappropriate rights. Think about what your identification solution will include. Do you use remote devices, bare metal, and the cloud, or do you rely on cloud-based apps? Determine the issues that IAM aims to resolve, then make those resolutions the main emphasis of your project plan.

2. Map the workforce to assign privileges

Determining who must have access to what resources is essential when starting an identity security project. To create a picture of each position and person within the company, as well as contractors or freelancers who need safe access, speak with HR.

Create continuing connections between the HR and security departments to allow privileges to be updated as new hires join the organisation, depart, or take on different responsibilities. To serve as a foundation for providing rights, it can also be required to compile a list of all linked devices, databases, and apps. Before altering any access procedures, this inventory creation activity can also serve as an audit to identify any outdated hardware or software updates.

3. Create individual profiles and intelligent role definitions

Each person who wants to access the network has to have a profile that lists all of their privileges. Don't give departments or contracting businesses privileges. Adopt a thorough strategy for managing privileged access that offers comprehensive details.

However, it is frequently impractical to continuously update each person's access privileges. Establish role-based rights and allocate those responsibilities to people as needed to simplify things. Teams that collaborate across network resources might benefit from the flexibility that role-based access control solutions provide by sometimes granting rights for predetermined durations.

4.   Adopt Zero Trust Network Access as standard

The best access management system for allocating privileges is ZTNA. Until their access credentials are verified, all users are viewed as suspect under the Zero Trust approach. There are no exceptions, not even for users at the executive level.

Any lateral movement via the network should be governed by the least privilege principle, which states that users should only be able to access the resources necessary for their function. In any situation, managers should be careful not to provide too many permissions.

5.   Make multi-factor authentication universal

One of the main components of safe IAM is the authentication procedure. Generally speaking, don't depend just on passwords to protect your data. Instead, incorporate a multi-factor authentication mechanism into user access portals.

MFA entails asking for one or more extra login credentials before allowing access. These credentials may consist of social media account authentication, codes given via email or SMS, or biometrics. An additional degree of security may be added by easily integrating third-party multi- or two-factor authentication providers into access management systems.

Since password-free access is becoming more and more popular, it would also be beneficial to think about doing away with passwords entirely. Include robust password security procedures in your IAM system at every level if this isn't a possibility.

6. Create centralized network visualization

Network managers have full visibility thanks to strong IAM solutions. They must be able to keep an eye on all endpoints, users, and activities within the network perimeter, including bare-metal and cloud devices.

In complicated organisations with several cloud databases or core programs, visibility is essential. For example, a business could have to link corporate accounts and human resources (HR) with eCommerce APIs and customer identity and access management (CIAM) systems. In these circumstances, cloud access management solutions—such as customer identity and access management solutions—will often be the best option.

Make sure that each user's device, location, and department are connected to the centralised IAM system. Centralization facilitates seamless onboarding and maintenance of orphaned accounts and improves the efficacy of real-time user access monitoring.

Additionally, if feasible, use single sign-on (SSO) procedures. Users ought to use a single set of credentials to access a single data point.

7. Audit orphaned accounts regularly

Employees' network identities may not always disappear when they leave a company. Hackers may use social engineering tactics to attack so-called orphaned accounts.

A Varonis investigation from 2021 found that over 10,000 of these phantom users were present in almost 40% of financial sector businesses. When individuals change positions or move on, it is crucial to locate orphaned accounts. To guarantee that orphaned accounts are quickly neutralized, schedule routine user management audits. Don't forget to include any partners or contractors who have left the firm.

8. Use automation to your advantage

Implementing onboarding features for customers or employees may significantly lower the labour and expense related to IAM.

Security personnel often do not need to customize individual access credentials during onboarding, and pre-assigned responsibilities can facilitate the seamless integration of new hires into network security processes.

Additionally, automate continuous password management. Self-service password portals, for example, can encourage staff members to strengthen their password security while lightening the strain for security teams.

The same is true for offboarding, where issues with orphaned accounts may be resolved with automation. Nonetheless, evaluate automated procedures on a frequent basis since responsibilities will need to be updated to account for evolving conditions.

9. Build IAM around regulatory compliance

One essential element of contemporary cybersecurity requirements is access management, and successful compliance will be facilitated by a strong IAM implementation.

Access control is necessary whether you're trying to comply with laws like the Payment Card Industry Data Security Standard (PCI-DSS), the Health Insurance Portability and Accountability Act (HIPAA), or the General Data Protection Regulation (GDPR) of the European Union.

Make sure you are completely covered by sector-specific regulations and include compliance into your planning from the start. Legal compliance isn't the only concern. Additionally, it's a great technique to focus project managers and establish credibility with clients or partners.

Identity and Access Management Risks in Business

IAM initiatives may go wrong in a lot of ways, therefore it's important to evaluate the main risks before starting any kind of access management change.

- **Management buy-in:** Every step requires the support of executives. The process should ideally have an executive advocate who can drive necessary structural or cultural changes to corporate processes or acquire financing and human resources.

- **Stakeholder engagement:** IAM should never be limited to a security issue. It is crucial to visualise the process as enterprise-wide while taking into account the requirements of various departments and users. It's also critical to communicate. Make sure the organisation as a whole and all-important stakeholder are aware of the project's objectives.

- **Procurement:** An IAM project may initially be hampered by improper technology sourcing. Although security teams' technical know-how must be relied upon, procurement choices should be made with a wider view in mind. When necessary, include professional guidance; nevertheless, before making any judgements, evaluate the business's needs and evaluate various goods.

- **Scale and strategy:** Over time, IAM systems may malfunction, particularly in the event of unforeseen modifications. Systems for managing access must expand with the business and collaborate well with outside contractors. Additionally, they should not be overly stiff and be adaptable to new technologies.

- **Policies and people:** Access management is both a technological endeavour and a continuous human struggle. Hearing what various departments and individuals have to say about their access requirements is crucial. Do their privileges suffice? Is IAM compromising its methods of operation? Additionally, teams must be established to audit access systems in order to guarantee security and efficiency.

- **Project drift:** Delivery might be delayed by inadequate management, as is the case with all IT projects. In general, it makes appropriate to map progress by breaking the project up into smaller parts with brief deadlines and milestones.

Having proper IAM habits is the only way to protect cloud environments, especially if the company works with AI and has a lot of valuable data or models. Other IAM strategies are the use of the principle of least privilege (PoLP) to limit access to an organization's data assets to only those resources needed by users in their roles; thus, there is less chance of data leakage. The second type is RBAC, which also enhances security by providing permissions based on roles in an organization; the third is ABAC, which also improves security by granting permission based on some attribute relevant to an entity. Biannual check-ups of IAM policies complemented by tools that track users' access and monitor for suspicious activity prevent compliance slips and strengthen security. Also, multi-factor authentication (MFA) increases security because it involves the use of multiple means of identification, making the work of the intruder much harder.

4.6 Cloud-Native Security Tools and Frameworks

The term "cloud-native security" refers to a security methodology that takes precautions to guarantee security at every stage of the unique lifecycle of cloud-native applications, from infrastructure planning to client delivery and maintenance. Every company has a policy on security. The majority of policies support having a system that is completely patched and impenetrable, and they oppose altering the configuration since doing so might leave some security flow.

The security situation for infrastructure today, however, is very different. It must move quickly and make adjustments. A completely safe organisation must undergo constant improvement and modification.

Rotate, Repave, and Repair are the three Rs of enterprise security that organisations must adhere to while automating infrastructure and providing continuous delivery.

- Rotate the stack credentials every hour or a few minutes.

- Every few hours, restart each server and application from a known good state.

- Fix susceptible application stacks and operating systems reliably within hours of a patch being available.

Source: - *(Gill, 2024)*

Common Types of Cloud-Native Security Solutions Available

- **Encrypted Data:** Several techniques are used to encrypt data while it is in transit and at rest in order to guard against illegal access and stop data leaks.

- **Network Security:** To guarantee safe data transfer, network security entails separating networks, managing access, and guarding against outside threats.

- **Security Scans:** Regular security scans, using both open-source and commercial tools, detect vulnerabilities in cloud infrastructure and applications.

- **Disaster Recovery Policy:** A disaster recovery policy outlines processes for restoring data and services after events such as natural disasters or system failures.

- **Securing DevOps with Cloud Native Security:** By integrating cloud-native security into DevOps, including security automation, patch management, and access control, organizations can reduce vulnerabilities and speed up secure software delivery.

Key Components of Cloud-Native Security Architecture Explained

These days, many companies are creating their own cloud-native security frameworks. For instance, Google's security solution Beyond Prod addresses development workflows, microservices, infrastructure, and pods. IBM, on the other hand, offers its own cloud-native secure infrastructure offering that supports network security, memory in-use protection, secure storage, and trusted computing features.

Different organizations in the market have different frameworks and policies to secure their environments. Some are designed as commercial frameworks to acquire more clients, while others are shared as open-source alternatives. However, in the end, an organization's security framework always depends on its business needs. Organizations without cyber security expertise tend to buy commercial or open-source security frameworks for their use, while those with cyber security expertise create their frameworks.

Identifying Threats to Cloud-Native Applications and Services

1. *Unauthorized Access:* Unauthorized access often occurs due to unsecured APIs or exposed legacy features. Cloud-native security platforms can help prevent such breaches by enforcing strong authentication and authorization controls across applications and resources.

2. ***Absence of Multifactor Authentication:*** Without multi-factor authentication (MFA), compromised credentials increase the risk of attacks, especially in cloud-native environments. Attackers find it more difficult to obtain unauthorized access when MFA is implemented because it offers an extra degree of protection.

3. ***Misconfiguration:*** Misconfiguration of cloud environments, such as using default settings or credentials, leaves systems vulnerable to breaches. Organizations must configure their cloud-based platforms properly and remove default settings to reduce attack surfaces.

4. ***Lack of Control:*** Using third-party cloud infrastructure services means losing control over security. Organizations should seek vendors offering cloud-native security services or on-premises solutions to maintain better control over security incidents and response times.

5. ***Data Privacy Concerns:*** Cloud vendors have administrative access, which can raise concerns over data privacy. To protect sensitive data, organizations must implement Zero Trust policies and audit logs and restrict data access through security controls.

How to block advanced persistent threats?

Armed attacks against targets intended to obtain data and important information instead of harming the organisation are known as advanced persistent threats. This assault stays hidden for a long time, gains a quiet understanding of how the entire stack functions, and eventually gains access to private information. Understanding the attack's mechanism will help us prevent it. Three factors are necessary for an attacker to initiate an assault.

- **Time:** APTs unfold over a long period, allowing attackers to gather information and act undetected.

- **Leaked Credentials:** Stolen login details enable unauthorized access to systems and data.

- **Unpatched Software:** Exploiting known vulnerabilities in outdated software gives attackers entry without detection.

What are the 3 R's of Cloud Native Security?

The method for making cloud-native settings secure is known as the "three R's of security." The fundamental idea behind the Three Rs of the Enterprise Security model is that assaults have a greater chance of causing significant harm the longer you allow them to continue. Therefore, it is better to accept the change and act swiftly. Let's take a closer look at the three Rs.

Rotate: Every few minutes, the data center's credentials for people, data centers, automated services, etc., should be changed. These credentials might be any kind of access token, password, or certificate. Although it's not always possible to prevent credentials from being leaked, changing them every few hours or minutes makes it more difficult for hackers to obtain them. This can be automated through a Cloud Native Security Platform to ensure security across cloud-based platforms.

Repave: Rebuild each data centre server and application from a known secure state. Destroying the outdated containers and virtual machines and reconstructing them from a known secure state is another way to fix the entire stack rather than just fixing the specific applications. This approach ensures the cloud-native security of Kubernetes security and containers, helping to avoid issues caused by unpatched software and maintaining overall cloud security.

Repair: Although vulnerable components should be repaved, securing the system from vulnerability should be prioritized. Therefore, whenever a vulnerability is found, the system, program, or method should be repaired as soon as possible. This helps make the system more secure by repairing the vulnerability and reducing the attack surface area. Implementing Cloud-Native Application Security best practices ensures any vulnerabilities in the stack are patched quickly, improving cloud infrastructure security.

Comparing Traditional Security with the 3 R's Approach

Security is currently the biggest issue with computer systems. Conventional methods of organisational security frequently impede progress and slow down development. Organizations have to set monitoring sensors and systems in place to check whether a security breach has happened or not. It is considered a reactive approach as detecting the threat is prioritized instead of vulnerability resolution. Patches are also applied step-wise to resolve the vulnerabilities at a later stage. This approach follows a methodology that is resistant to modern technologies. We are aware that the likelihood of possible harm increases with the amount of time the attacker has to breach the system.

Whereas its security three R's provide an automated vulnerability resolution mechanism. It uses a proactive approach to change the system configurations efficiently. The vulnerabilities cannot be replicated, and the virus or worm would be removed as soon as possible. Instead of patching, the vulnerable components are created from scratch, which is modelled to reduce the vulnerability from the beginning. The three R's are faster, better, and more secure than traditional approaches; using them with modern technologies is very easy. As a result, the 3 R model has altered the perception of cloud native security.

Tradition Security	3 R's of Security
Monitored and Instrumented Systems - Businesses set up monitoring to identify any changes if there is a security breach.	**Automated** - The system must be upgraded as soon as possible. Immutable infrastructure and automation can assist in removing settings that compromise system security.

Tradition Security	3 R's of Security
Reactive - Priority one should be given to threat detection, followed by vulnerability further resolution.	**Proactive** - Changing the system's state is crucial in order to prevent the virus from reproducing and surviving.
Patched Incremental - To fix the problem, patches are gradually applied to the outdated system.	**Fresh, Clean State Deployment**—New clean images are utilised to deliver items automatically rather than fixing the outdated systems.
Resisting Changes - Patching outdated systems that are resistant to modify is its preferred method.	**Promoting Changes** - This method is secure and handles updates more quickly.

Source: - *(Gill, 2024)*

AI security solutions and architectures are necessary to ensure security in AI-driven cloud structure as the new security tools and strategies are developed to address inherent security risks in contemporary architectures. They include, but are not limited to features such as threat intelligence automation, continuous monitoring, and Infrastructure as Code scanning to detect threats in development stages. The CSA offers a matrix for deploying standard security controls across cloud solutions; on the same note, the NIST gives guidelines on the same. AWS Security Hub, Microsoft Defender for Cloud, and Google Security Command Centre are examples of native security solutions offered by cloud service providers that help businesses manage compliance, spot configuration errors, and automate actions. By using these cloud-native solutions, a proactive security posture that complies with legal requirements for safeguarding data, workloads, and AI models in the cloud environment is made possible.

4.7 Chapter Summary

This chapter discussed primary cloud security and compliance issues in AI solutions and focused on how to prevent adverse impacts on data, models, and related structures. The threats were introduced within the context of cloud security architecture and how threat should be managed proactively. The chapter also discussed data privacy and compliance issues, focusing on following the standard of protection for sensitive information worldwide. The secure AI workflows were considered with emphasis on the cryptographic protection, access control measures, and the secure model deployment. Key IAM practices were described in relation to the principle of least privilege, the use of MFA, and, finally, the constant review of access rights. Last but not least, the chapter described the newest approach to cloud security – cloud-native security tools and frameworks, including AWS Security Hub and compliance with NIST guidelines for protecting AI-driven cloud applications.

Multiple Choice Questions (MCQs)

1. **Which of the following is a primary goal of cloud security architecture?**

    a. Enhancing user interface design

    b. Mitigating security threats in cloud environments

    c. Optimizing cloud storage for large datasets

    d. Reducing cloud service costs

2. **What is the primary purpose of Identity and Access Management (IAM) in cloud security?**

    a. Monitoring cloud storage performance

    b. Controlling user access to cloud resources

    c. Encrypting all cloud data automatically

    d. Managing software development workflows

3. **Which of the following is an example of a cloud-native security tool?**

    a. Kubernetes

    b. Virtual Private Network (VPN)

    c. Amazon Guard Duty

    d. HTTP Secure (HTTPS)

4. **Why is compliance critical in AI-driven cloud environments?**

    a. It improves cloud application performance.

    b. It ensures adherence to data privacy regulations.

    c. It enhances AI algorithm efficiency.

    d. It reduces operational costs.

5. **Which regulation is commonly associated with data privacy compliance in cloud environments?**

    a. ISO 9001

    b. GDPR

    c. ITIL

    d. PCI DSS

6. **What is a key characteristic of secure AI workflows in cloud environments?**

    a. Automation of encryption processes

    b. Scalability without performance impact

    c. End-to-end encryption of data

    d. Real-time error correction

7. **Which component is crucial for threat mitigation in cloud security architecture?**

    a. Multi-factor authentication

    b. Cloud billing optimization tools

    c. Low-latency data storage

    d. Enhanced network throughput

8. **What is the role of cloud-native security frameworks in AI-driven environments?**

    a. To reduce the cost of cloud resources

    b. To provide tools for managing AI training datasets

    c. To integrate security directly into cloud infrastructure

    d. To enable faster development of AI models

9. **Which IAM best practice helps prevent unauthorized access in cloud environments?**

    a. Assigning minimal privileges to users

    b. Using default usernames and passwords

    c. Sharing credentials across teams

    d. Disabling encryption to reduce complexity

10. **What is a key benefit of implementing cloud security measures in AI workflows?**

    a. Increased computational speed

    b. Enhanced protection against cyberattacks

    c. Reduced complexity of AI algorithms

    d. Simplified cloud billing processes

Answer

1	2	3	4	5	6	7	8	9	10
b	b	c	b	b	c	a	c	a	b

Chapter 05

SCALABILITY AND PERFORMANCE OPTIMIZATION IN CLOUD-AI SYSTEMS

5.1 Chapter Overview

This chapter specifically deals with the discussion of scalability and performance techniques of the cloud AI system. With a greater reliance on AI solutions in organizations, it becomes necessary to ensure that these systems can scale up for both data and users as well as computation. The first section of the chapter provides the reader with information about what is meant by scalability within cloud-AI systems and illustrates how vertical and horizontal scaling work. It proceeds to discuss about effectiveness of the system in relation to resource management, system utilization and data handling techniques. Other factors like latency, failure, and cost are also explained to assist various organizations in sustaining the high availability and response time in AI related operations. Further, the chapter also focus on auto-scaling, which together with the container orchestration and distributed computing frameworks help in achieving scalability and performance at the same time. At the end of this chapter, the readers get acquainted with important information on possible adjustments of the cloud-AI systems and their optimal design to meet new requirements in terms of performance and reliability.

5.2 Designing Scalable Cloud Architectures

Cloud computing Scalability is the capacity of cloud-hosted systems to adapt to shifting demands by increasing or decreasing their "capacity." Convenience, flexibility, cost savings, and enhanced disaster recovery are some advantages of cloud scalability. To put it another way, a scalable system saves money during lower-than-normal loads while adapting to manage higher traffic without overprovisioning.

Definition and Importance of Scalability

The capacity of a cloud architecture to scale up or down resources in response to shifting workload needs is known as scalability in cloud computing. Businesses may simply add or subtract computer resources as needed because to this flexibility, which eliminates the need for large infrastructure upgrades or hardware investments. A key component of cloud computing is scalability, which helps businesses swiftly adjust to changing needs and guarantee peak performance and client satisfaction. Businesses can maintain high levels of service availability and responsiveness by utilising scalable cloud infrastructure, which is crucial for being competitive in the rapidly evolving digital world of today.

Benefits of Scalability

Scalability in cloud computing has several advantages that affect companies of all sizes.

First of all, scalability enables businesses to promptly adjust to shifting needs, guaranteeing that their services and apps continue to function well and be responsive even during periods of high traffic. Because users encounter less downtime and quick reaction times, this flexibility results in optimal performance and increased customer satisfaction.

Another important advantage of cloud computing scalability is cost savings. Businesses can avoid the costs of overprovisioning by only purchasing the resources that are truly required. Businesses may better manage their finances and direct resources where they are most needed with this pay-as-you-go arrangement.

Additionally, scalability allows companies to expand internationally and connect with clients in several locales without requiring physical infrastructure in each one. By providing scalable solutions that may be implemented across several geographical locations, cloud providers provide this global reach and guarantee constant performance and availability for consumers everywhere.

In conclusion, the advantages of cloud computing scalability include the capacity to grow internationally, save costs through effective resource use, and promptly adjust to shifting demands—all of which enhance performance and customer happiness.

Cloud elasticity vs. cloud scalability

The capacity of a cloud environment to adjust the necessary resource allocation based on workload is known as cloud elasticity. A feature of the cloud called elasticity allows you to meet needs without overprovisioning resources, which lowers the amount of money you spend on cloud infrastructure.

On the other side, cloud scalability refers to the capacity to adjust resource allocation in response to fluctuating workloads. It is a feature of the cloud environment that ensures a scalable cloud environment by helping you keep your system's performance from declining even when demands may increase.

Types of scalability in cloud computing

There are several approaches to scaling in cloud computing, which basically involves altering the resources' capacity to manage traffic. There are three different kinds of scaling: diagonal, horizontal, and vertical.

1. **Vertical scaling**

   Another name for vertical scaling is scaling up or down. It is the simplest sort of scaling and ought to function immediately with any application. It involves allocating more CPU or RAM to each instance and making sure there is enough processing power without modifying your code or application.

Increasing the resources of instances that host your application essentially makes it more capable of providing content to a wider user base.

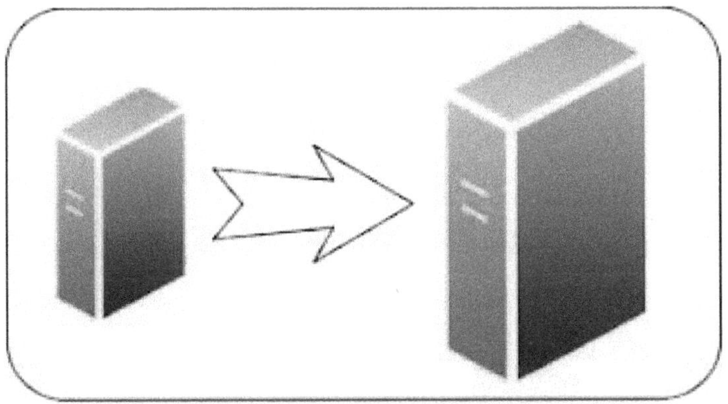

Vertical scaling

Source: - *(Cebula, 2024)*

2.   Horizontal scaling

Horizontal scaling conjures up images of scaling in or out. The goal here is to increase the number of instances operating in parallel while maintaining the same resource allocation for each instance.

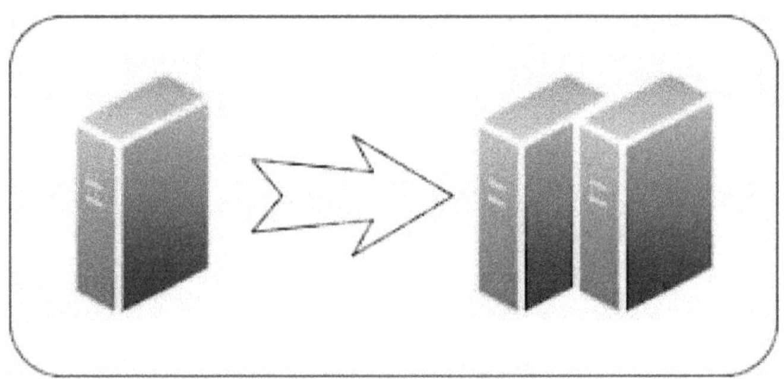

Horizontal scaling

Source: - *(Cebula, 2024)*

3. Diagonal scaling

Vertical and horizontal scaling are combined to create diagonal scaling. Your instances can be scaled up simultaneously with being scaled out to accommodate increased traffic. Furthermore, you may set up parameters to initiate scaling automatically in response to consumption levels, guaranteeing efficient resource management and avoiding performance deterioration.

This is helpful when a single process on a single instance starts using up the majority of its resources; in this case, the instance is scaled up, and scaling out creates smaller units to handle the remaining traffic.

5.2.1 Building Scalable Cloud Architecture

Scalable architecture should be considered from the very beginning of the infrastructure design process. Why? There are several benefits to scalable cloud computing for companies of all sizes.

First off, since you don't have to pay for resources you don't utilise, scalable infrastructure is economical. By utilising cloud services that are extremely scalable, you may create dependable solutions that can withstand increases in workload.

Additionally, cloud services are loaded with security measures since cloud providers prioritise infrastructure security. Furthermore, delaying the addition of scalability to your infrastructure is a dangerous move. Why?

Imagine that your application's user base is increasing surprisingly quickly and that your product launch is a big success. As a result, you have to remove the software and restructure your environment since your infrastructure is starting to get overburdened. These kinds of complicated circumstances might cost your company clients and revenue.

Designing for Scalability

To guarantee that a cloud infrastructure can manage rising workload needs without sacrificing performance, scalability design is crucial. This entails creating and putting into place a scalable architecture that can change to meet evolving needs. By taking scalability into account early on, companies can create a solid cloud architecture that facilitates expansion and guarantees steady performance.

- *Modularity*

  Since your application is where it all begins, we advise using a modular design. More manageable, safe, and scalable than monolith apps, microservices (a modular approach) divide large components into simpler, smaller bits.

- *Containerization*

  Containerize your application if it is modular. By taking use of third-party cloud providers' capabilities, you can make sure your application has the resources and infrastructure it needs to grow effectively. For instance, you may manage cloud scalability from a single control panel when you run your application using tools like Docker or Kubernetes. Additionally, we suggest that you keep your app image as small as possible and make your application stateless. It facilitates faster, simpler, and undetectable scaling.

- ***Designing for Scalability from the Start***

  Planning the cloud infrastructure for scalability from the beginning entails taking an organization's possible development into account. This entails selecting a cloud provider that provides scalable solutions, putting in place a scalable architecture that can change to meet evolving needs, and utilising load balancing and auto-scaling to guarantee peak performance. Businesses may build a cloud infrastructure that is ready for future expansion and can manage rising workload needs without sacrificing performance by following these steps.

- ***Using Auto-Scaling and Load-balancing***

  To achieve scalability in cloud computing, load balancing and auto-scaling are crucial techniques. By dynamically adding or removing computing resources in response to workload needs, auto-scaling enables organisations to guarantee that the proper quantity of resources is always accessible. By just purchasing the resources that are required, this lowers expenses and helps avoid overprovisioning.

  In contrast, load balancing makes sure that incoming traffic is split equally across several servers, avoiding any one server from experiencing overload. This increases application and service availability and dependability in addition to performance.

  Organisations may guarantee that their cloud infrastructure is scalable and capable of managing rising workload needs without sacrificing performance by implementing load balancing and auto-scaling. By effectively managing computer resources, this helps businesses lower costs, enhance performance and customer happiness, and swiftly adjust to shifting needs.

  All things considered, scalability in cloud computing and the capacity of businesses to promptly adjust to shifting needs while guaranteeing peak performance and client delight depend on planning for scalability and using auto-scaling and load balancing.

- ***Reliability***

  Most of the time, cloud services provide cross-region replication and are by default highly available. Utilising those services reduces downtime and increases the resilience and dependability of your application. After all, you'll probably lose some clients' confidence and have to say goodbye to them if you're unavailable, which will also cost you money.

- ***Monitoring***

  This is a crucial component of automation as it helps you choose what can and should be automated, reducing human error and boosting speed and consistency. Monitoring infrastructure metrics also aids in managing data storage capacity, guaranteeing that your system can effectively expand to accommodate changing or expanding demands. Additionally, you may make data-driven scaling decisions by keeping an eye on infrastructure metrics and setting up warnings for unusual conditions.

- ***Automation***

  Autoscaling enables you to grow resources in response to increased demand without the need for engineers, it simplifies lives. You only utilize the server power you require, when you require it, thanks to auto-scaling, which is based on tracked data and is a convenient cost-management tool. Additionally, continuous integration and continuous delivery (CICD), which entails quicker and automated application deployment, should be taken into account.

- ***Security***

  Security is essential; your company suffers if you can't safeguard your services. Therefore, it is essential to include security in the original design. Numerous security safeguards are integrated into cloud services. For instance, cloud computing providers that priorities data security, such as AWS, offer CloudTrail, security groups, and identity access management (IAM).

AWS Services to Build Scalable Computing Environments

One of the primary benefits that cloud environments have over on-premise ones is scalability. Businesses can swiftly grow apps and services without the drawn-out setup time that comes with traditional equipment thanks to cloud solutions that do away with the requirement for actual hardware. You may employ a number of cloud-native services that offer excellent scalability and dependability right out of the box, such as:

- **ASG (Auto Scaling Group):** This is just a collection of EC2 virtual machines and the most straightforward method of giving your application scalability. You may scale the environment using your own custom CloudWatch alarms or preset metrics like CPU or network utilisation by simply preparing a machine image (AMI) with your application and configuring the scaling policies.

- **ECS (Elastic Container Service):** This fully managed container orchestration solution facilitates the deployment, management, and scalability of containerized applications. It handles scaling your application containers and supporting infrastructure automatically.

- **EKS (Elastic Kubernetes Service):** Running Kubernetes on AWS and on-premise is made simple with this managed Kubernetes solution. It also aids in the construction of your Kubernetes cluster.

- **S3 (Simple Storage Service):** Files are stored on this highly accessible scalable storage service, which offers many levels for storage reliability and access frequency. Because of its pay-as-you-go implementation, data storage is never overprovisioned.

- **RDS (Relational Database Service):** The most widely used engines, including MySQL and PostgreSQL, may be used to set up, run, and scale relational databases using this managed service. A database instance can be easily scaled up, and RDS gives you the flexibility to dynamically adjust the amount of storage that is allotted as user space demands rise. Additionally, databases that include reader or writer replicas might be scaled out.

Cost Optimization Strategies in Cloud Scalability

One of the key benefits of scalable cloud architecture is the potential for cost reductions, but using this requires a strategic approach. Businesses may reduce their cloud costs by selecting the right pricing models, including AWS's Reserved Instances and Spot Instances. It is crucial to efficiently manage increased storage capacity in addition to CPU power and networking resources in order to manage unpredictable loads and achieve performance improvements.

Reserved instances, which are perfect for workloads that are predictable, let businesses commit to cloud resources at a reduced cost. However, for flexible workloads where resources can be scaled down without affecting the company, Spot Instances provide considerable cost reductions. Businesses may cut costs and prevent overprovisioning by dynamically modifying cloud capacity in response to variations in workload and traffic.

Additionally, by only billing for resources consumed, the pay-as-you-go approach improves cost management even more. Businesses must carefully weigh this against the possibility of under-provisioning, though, as this might result in performance snags during periods of high traffic. Combining these tactics guarantees that companies may get peak performance while controlling cloud expenditures.

Building for the public cloud requires engineering solutions that can accommodate various levels of utilization while keeping them performant and still affordable. The strategies used are correlated with horizontal scaling techniques, where one adds more instances to the setup to share the workload, and vertical scaling, where one enhances the capability of existing resources. Load balancers, auto-scaling groups, and the container orchestration platform – Kubernetes – are the kinds of cloud-native tools that help to deal with dynamic workloads. Managing resources, preparing for potential failures, and using microservices to build applications also improve scalability by enabling the scaling of services separately. Therefore, when designing, the organizations should consider scalability to maintain top level of system performance, high availability and user experience, especially during times of peak traffic and usage or especially during the execution of intensive AI processes.

5.3 Elasticity and Load Balancing in AI Workflows

One method for distributing network traffic among a group of computers called a server farm is load balancing. By distributing demand evenly among several servers and computing resources, it maximizes network performance, dependability, and capacity while decreasing latency.

In order to prevent excessive network traffic from overwhelming a single server, load balancing employs a physical or virtual appliance to determine in real time which server in a pool can best fulfil a certain client request.

In addition to maximizing network capacity and ensuring high performance, load balancing provides failover. If one server fails, a load balancer immediately redirects its workloads to a backup server, thus mitigating the effect on end users.

The Open Systems Interconnection (OSI) communication model's Layer 4 and Layer 7 are typically supported by load balancing. Traffic is distributed by Layer 4 load balancers using transport information like IP addresses and TCP port numbers. Application-level features, such as Hypertext Transfer Protocol (HTTP) header data and the message's actual contents, such URLs and cookies, are used by Layer 7 load-balancing devices to determine routing. Although Layer 4 load balancers are still widely used, especially in edge installations, Layer 7 load balancers are increasingly prevalent.

How load balancing works

Users' requests for information and other services are handled by load balancers. They are situated between the internet and the servers that process such requests. The load balancer directs a request to the server in a pool that is available and online after first identifying which server is online. A load balancer may dynamically add servers in response to traffic surges and responds quickly during periods of high load. On the other hand, if demand is minimal, load balancers may drop servers.

How load balancing works

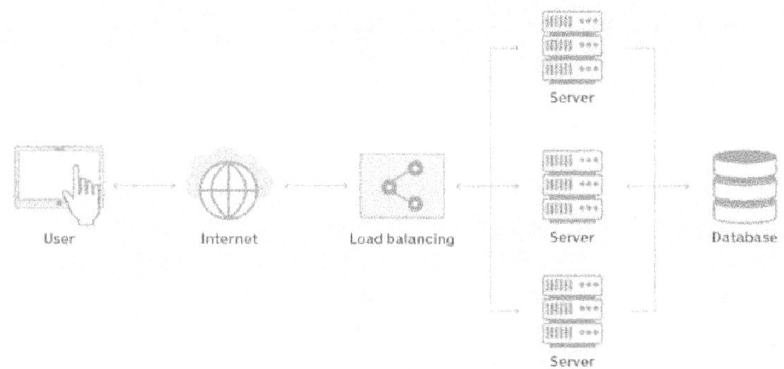

Source: - *(Yasar, 2024)*

Types of load balancers

One essential element of highly available infrastructures is load balancing. Different load balancer types with varying storage capacities, features, and complexity can be implemented based on a network's requirements.

A load balancer may be a software program, a hardware device, or a mix of the two. Two varieties of load balancers are as follows:

1. **Hardware load balancer.** A hardware device with proprietary, specialised software built in that can manage enormous volumes of application traffic is called a hardware load balancer. Because of their integrated virtualization feature, these load balancers allow several virtual load balancer instances to run on a single device.

   Vendors used to sell consumers standalone appliances with proprietary software installed onto specialised hardware, typically in pairs to offer failover in the event that one system failed. Expanding networks need that a company buy more or bigger appliances.

2. **Software load balancer.** Virtual machines (VMs) or white box servers are used for software load balancing, which is most often an application delivery controller (ADC) feature. Additional functionality like caching, compression, and traffic shaping are frequently provided by ADCs. Virtual load balancing, which is common in cloud systems, may provide a great deal of flexibility. For instance, it allows users to automatically scale up or down in response to drops in network activity or surges in traffic.

Cloud-Based Load Balancing

Cloud load balancing balances cloud computing settings by using the cloud as its underlying infrastructure.

Examples of cloud-based load-balancing models include the following:

- **Network load balancing:** The quickest load-balancing solution is this one. It transports network traffic using network layer information and runs on Layer 4 of the OSI architecture.

- **HTTP Secure load balancing:** This makes it possible for network managers to allocate traffic according to the data originating from the HTTP address. It is one of the most adaptable load-balancing solutions and is based on Layer 7.

- **Internal load balancing:** Although it may also balance traffic distribution within the internal infrastructure, this is comparable to network load balancing.

Load-balancing algorithms

The servers that receive particular incoming client requests are chosen by load-balancing algorithms. Static and dynamic load balancing algorithms are the two primary categories.

1. Static load-balancing algorithms

- The IP hash-based method uses specified keys, including HTTP headers or IP address data, to determine a client's preferred server. Applications that depend on user-specific saved state information, like e-commerce checkout carts, benefit from this technique's support for session persistence, also known as stickiness.

- The round-robin technique uses the domain name system (DNS) to distribute traffic to a list of servers in rotation after going through all of the accessible servers in sequential order. A list of several "A" records is carried by an authoritative nameserver, which responds to each DNS query with one.

- Administrators can give each server a different weight by using the weighted round-robin technique. In this manner, depending on their weight, the servers that can manage higher traffic volumes get a little more traffic. DNS records are where weighting is configured.

2. *Dynamic load-balancing algorithms*

- In addition to checking and sending traffic to servers with the fewest open connections, the least-connections technique gives preference to those with the fewest active transactions. This algorithm makes the assumption that each connection needs roughly the same amount of processing power.

- The weighted least connection approach makes the assumption that certain servers are more capable of managing higher traffic volumes than others. As a result, administrators can give each server a varied weight.

- In order to determine the optimal destination for traffic transmission, the weighted response time technique combines the average response time of each server with the number of active connections. Because it routes traffic to the servers with the fastest response times, this algorithm guarantees speedier service.

- The resource-based approach allocates load according to each server's current resource availability. Prior to traffic distribution, it checks the availability of the CPU and RAM using specialised software known as an agent that runs on each server.

Benefits of load balancing

Network traffic load-balancing is quite advantageous for organisations that oversee several servers. These are the primary benefits of employing load balancers:

- **Improved scalability:** Depending on the needs of the network, load balancers may grow the server infrastructure as needed without compromising services. For instance, a website may have an abrupt increase in traffic if it begins to draw a lot of visitors. The website may crash if the web server cannot handle this unexpected spike in traffic. To avoid this, load balancing can distribute the additional traffic among several servers.

- **Improved efficiency:** Response times are improved and network traffic flows more smoothly as a result of each server having less load to handle. In the end, this gives site visitors a better experience.

- **Reduced downtime:** Load balancing can help businesses with various sites in different time zones and a worldwide presence, particularly with server maintenance. To avoid service disruptions or downtime, a business might, for instance, shut down the server that requires repair and redirect traffic to the other load balancers that are accessible.

- **Predictive analysis:** Early failure identification and management are made possible by load balancing, which doesn't interfere with other resources. For instance, load balancers that are software-based can anticipate traffic jams before they occur.

- **Efficient failure management:** In the case of a breakdown, load balancers have the ability to immediately reroute traffic to backup choices and operational resources. For example, load balancers can reallocate resources to other unaffected locations in order to reduce service disruption if a failure is discovered on a network resource, such a mail server.

- **Improved security:** Load balancers increase security without requiring more resources or modifications. The offloading functionality is one of the security features that load balancers are being equipped with as more computing shifts to the cloud. This protects a company from distributed denial-of-service assaults by moving attack traffic from the company server to a cloud service provider.

Hardware vs. software load balancers

Pros

- Since specialised processors are used to operate the program, they offer quick throughput.

- These load balancers provide enhanced security since the company handles them alone, without the involvement of any other parties.

- They have a set price at the moment of purchasing.

Cons

- Configuring and programming hardware load balancers requires additional personnel and knowledge.

- Once a certain number of connections has been achieved, they are unable to scale. Connections are either denied, dropped, or deteriorated as a result, and installing more computers is the only way to fix this.

- They cost more because they are more expensive to buy and maintain. If you own a hardware load balancer, you may need to hire specialists to handle it.

Software load balancers

Pros

- They provide the adaptability to change to meet a network's evolving demands and specifications.

- They can expand beyond the initial capacity by adding new software instances.

- They give off-site choices that can run on an elastic network of servers using cloud-based load balancing. Additionally, cloud computing provides a range of combinations, including hybrid with on-site sites. For instance, a business may have a backup load balancer in the cloud and the primary load balancer on-site.

Cons

- The first delay may be caused by software load balancers when growing over capacity. When the load-balancer software is being setup, this often occurs.

- Since software load balancers don't have a set initial cost, updates may become necessary over time.

Load balancing and elasticity are among the core principles with regard to the AI operations within the cloud environment in order to achieve the optimal usage of resources as well as consistently high performance of systems. Elasticity simply means the characteristic that the service can flexibly increase or decrease the amount of resources available to it. In the context of the AI workflows, this becomes even more critical because of the unpredictability of the nature of data processing and model training tasks that can include simple inference tasks through complex model training tasks that require significant computational resources. Elasticity is achieved through auto-scaling features that define, for cloud platforms, the ability to provision or release resources based on demand, while optimizing the cost-effectiveness and efficiency of cloud services.

The feature complements elasticity by serving to distribute the incoming workloads on available computing instances and avoid situations which may result in utilization of the available resources. In AI, load balancing is important when working with tasks such as, real time data inference, batch processing and distributed training tasks where the tasks have to be divided across multiple nodes. A load balancer is a tool that divides work and controls traffic so that no single server becomes overwhelmed, and there are such as Elastic Load Balancer (ELB) in AWS, Azure Load Balancer, and Google Cloud Load Balancing.

Incorporation of the elasticity feature together with load balancing simplifies the enhancement of a system's ability to recover from failures. For example, in case of a node containing the AI computations, the load balancers are capable of rerouting traffic to other working nodes while the infrastructure becomes elastic to bring up new nodes. This approach entails high availability and continuity, which is desirable for real-time AI applications such as setting up an AI-powered chatbot support system and fraud detection systems. Also, both strategies bring increased security since workloads are isolated and cannot be attacked by a DoS attack since traffic is well distributed.

Finally, the focus on elasticity and load balance approaches can be regarded as the key to making the AI workflows running in the cloud as scalable and effective as possible. Through these practices, an organization can manage increasing big data, varying workloads, and high complexity computations without making system availability or cost prohibitive. These capabilities ensure that there is constant improvement in the design and implementation of AI while at the same time operation is running and resources are well utilized.

5.4 Performance Monitoring and Optimization Tools

The function of performance monitors of optimization of the processes can be very important for the effective, reliable and safe usage of the

AI operations in cloud resources. These tools assist an organization to monitor KPI's of latency, throughput, resource utilization, and error rates in order to prevent performance issues that may hinder the organization. If the businesses are consistent on the status of the infrastructure, it will be easier to deploy resources and prevent unnecessary downtime which is vital for the efficient running of the artificial intelligence models and data flows.

Optimization tools work in conjunction with monitoring to offer specific solutions and the automation of functions that improve such systems. A notable aspect of the system is that, it self-scales, identifies anomalies in the application's performance and incorporates predictive analysis which ensures optimization of performance to a level best while at the same time ensuring that the costs are low. Some of the most common cloud solutions developed specifically for cloud-native applications are AWS CloudWatch, Azure Monitor, and Google Cloud Operations Suite, which contain effective dashboards and notification for real-time and past performance. Applying these tools helps to keep AI based systems performant, adjustable to the new loads and ready to work in production environments.

Cloud Monitoring Tools to Ensure Optimal Cloud Performance and Drive Business Success

Making sure your cloud infrastructure runs smoothly is not only essential, but also a competitive advantage. Monitoring performance, identifying problems, and doing real-time troubleshooting are more crucial than ever as more and more firms shift their operations to the cloud. Cloud monitoring solutions are useful in this situation.

In the cloud, cloud monitoring technologies act as your company's eyes and ears. They provide real-time insight into the cloud-based services, apps, and infrastructure you have. Along with monitoring and supplying useful data, they also assist in trend analysis, performance evaluation, and the early detection of any issues before they have an influence on your business operations.

Cloud monitoring solutions come in a variety of forms, but they are not all made equal. By giving insight into resource usage, performance levels, and any problems, the finest cloud monitoring tools help businesses maintain operational efficiency and manage their cloud environments. This article explores the world of cloud monitoring tools, presenting 11 of the best ones available right now and describing what to look for in those tools.

How to assess a cloud monitoring tool for your business

The difference between keeping ahead of problems and always having to catch up might be found in the choice of cloud monitoring tool. It all comes down to picking a solution that meets your company's requirements, complements your cloud monitoring plan, and provides the ideal ratio of functionality, usability, and scalability.

The top cloud monitoring solutions can provide you a thorough picture of your cloud environment, enhance decision-making, boost productivity, and eventually save expenses. Choose a product that best fits the demands of your business, keeping in mind aspects like cost, scalability, usability, and compatibility with external tools.

These are some essential characteristics to consider when selecting a new cloud monitoring solution:

- **Comprehensiveness of monitoring:** A thorough, real-time picture of your complete cloud infrastructure, including servers, databases, apps, and network performance, is what you should look for.

- **Integration and compatibility:** The application should be able to easily interact with other systems inside your company and work with your cloud platforms.

- **Scalability and flexibility:** It should be possible for your cloud monitoring solution to grow with your company and adapt to your changing requirements without experiencing any performance issues. This is especially crucial if your architecture is hybrid or includes many cloud environments.

- **Alerting and reporting:** A decent solution should deliver automatic, customisable reports providing important insights into system health and any problems, as well as real-time warnings depending on your specified circumstances.

- **User-friendly interface:** To speed up decision-making, the interface should be simple to use and have customisable views and visual data representations.

- **Security features:** Select security threat monitoring technologies that may be able to interface with current security information and event management (SIEM) systems.

- **Cost management:** Some cloud monitoring systems provide functions to assist in tracking your cloud expenditures, including suggestions for cost reductions and warnings of any cost overruns.

- **Customer support and community:** Tools with a strong user base, frequent updates, and strong customer service may give helpful support and best practices when needed.

Cloud Service Provider Built-in Monitoring Tools

As an essential component of your cloud infrastructure monitoring, built-in cloud monitoring solutions are made to measure performance in their respective settings with ease. The primary advantage of utilising these solutions is their native integration, which enables efficient troubleshooting inside the particular cloud platform, excellent performance tracking, and ease of use.

1.   Digital Ocean Monitoring and Uptime

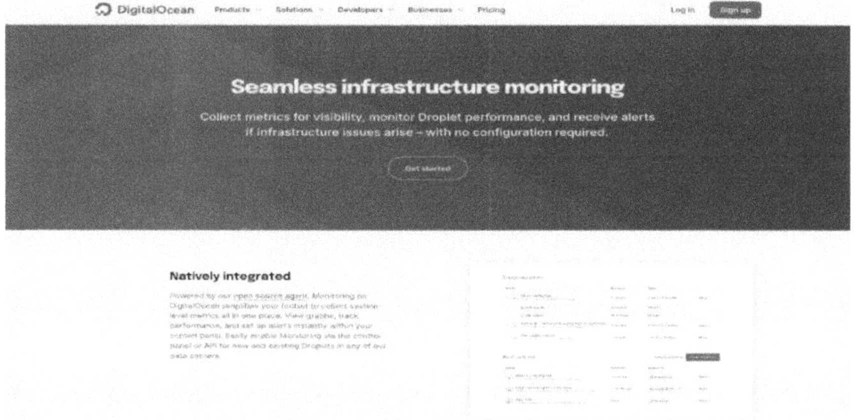

Source: - *(Ocean, 2024)*

Digital Ocean Monitoring is a free tool that gathers information about resource usage at the Droplet level for customers of Digital Ocean. With the option to design customized metrics alert settings and the ease of integrating email and Slack notifications, it provides improved Droplet visualisation tools that assist keep an eye on the health and performance of your infrastructure.

A service called Digital Ocean Uptime can assist you in keeping an eye on the latency and uptime of your websites and resources. With checks at one-minute intervals and customizable down to one millisecond of latency detection, it notifies you of endpoint problems throughout Digital Ocean's four worldwide zones. Uptime makes it easy to provide your users with the greatest possible app or website experience by allowing you to identify latency, outages, or impending SSL expiration.

2. AWS CloudWatch

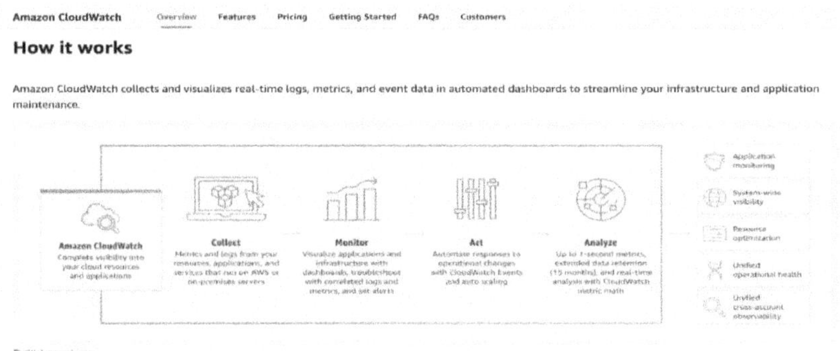

Source: - *(AWS, 2024)*

AWS CloudWatch provides a complete cloud monitoring solution for apps and resources in on-premises, Amazon Web Services, and other cloud environments. With its sophisticated visualisation tools, automatic alerts, connection with other AWS services, and dashboards that offer useful information, CloudWatch makes it easier to maintain your apps and infrastructure.

3. Microsoft Azure Monitor

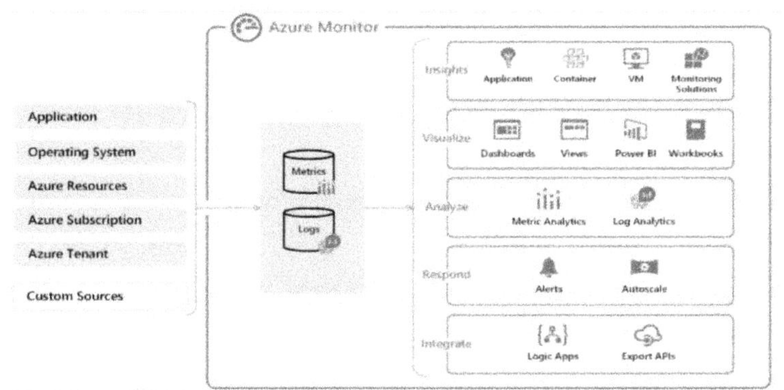

Source: - *(Challenge, 2024)*

In order to maximize the performance and availability of your apps and services, Microsoft Azure Monitor offers a comprehensive cloud monitoring solution that makes it possible to gather, analyze, and react to telemetry from on-premises and cloud environments. It collects data from all areas of your system, provides analysis and correlations using a shared set of tools, reacts automatically to system events, and integrates with tools and systems from third parties. Additionally, it allows custom sources and a variety of resource kinds via its APIs.

4. Google Cloud Operations

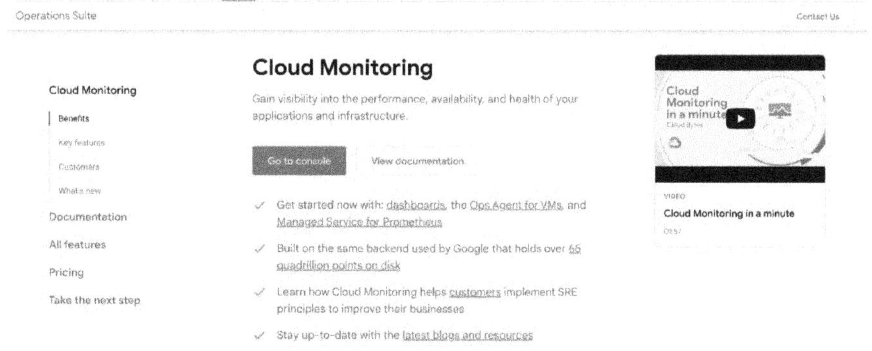

Source: - *(Google Cloud, 2024)*

An integrated solution for tracking, logging, and monitoring systems and apps on the Google Cloud platform and beyond is provided by Google Cloud Operations, formerly known as Stack driver. To enhance the performance, uptime, and general health of cloud-powered applications, features include real-time log management and analysis, metrics observability at scale, a standalone managed service for Prometheus, and Application Performance Management, which combines monitoring and troubleshooting capabilities.

Third-Party Cloud Monitoring Tools

A flexible solution that works with different cloud settings and platforms is provided by third-party cloud monitoring solutions. These tools are a great option for companies that operate in a variety of cloud settings because of their wide range of features, customisation options, and capacity to offer a comprehensive picture of your multi-cloud or hybrid cloud infrastructures.

1. Datadog

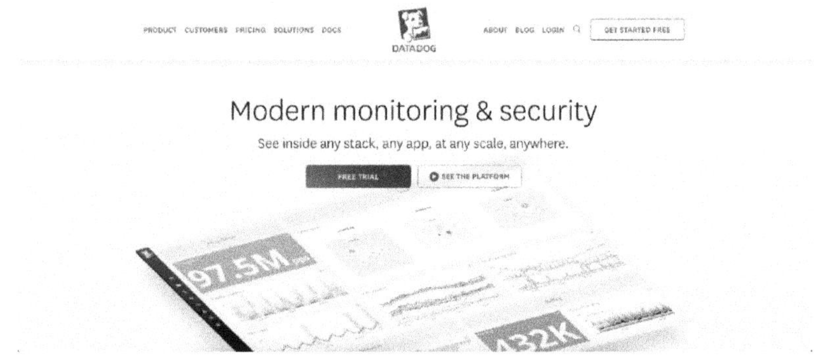

Source: - *(DataDog, 2024)*

A SaaS-based platform that gives comprehensive metrics, visualizations, and alarms to optimize cloud or hybrid settings is provided by Datadog's infrastructure performance monitoring. Real-time performance insights, tag-based analytics, machine learning-based alert tools, comprehensive technology coverage, sophisticated metric collection capabilities, and an easy-to-use interface that facilitates communication and troubleshooting—all of which lessen the need for costly professional services or extensive training.

2.   AppDynamics

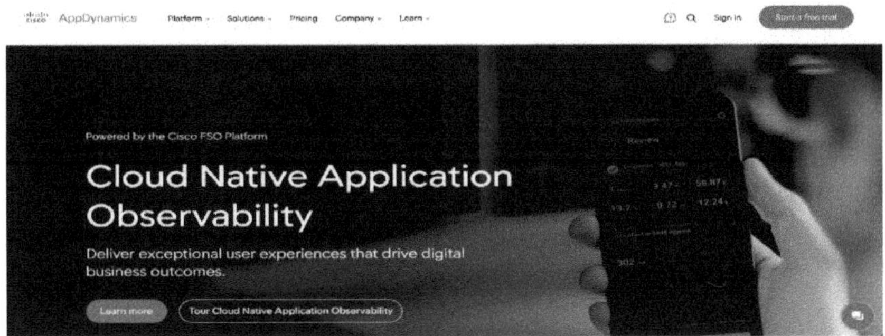

Source: - *(App Dynamics, 2024)*

AppDynamics is an all-inclusive cloud-native platform that emphasises application observability, guaranteeing that companies can deliver exceptional user experiences that complement their digital goals. With the help of the platform, users can visualise and monitor the performance of the complete technological stack, connecting technical data to business results. End users can quickly detect and fix any problems thanks to this full-stack viewpoint, reducing the possibility of negative effects on company performance.

3.   New Relic

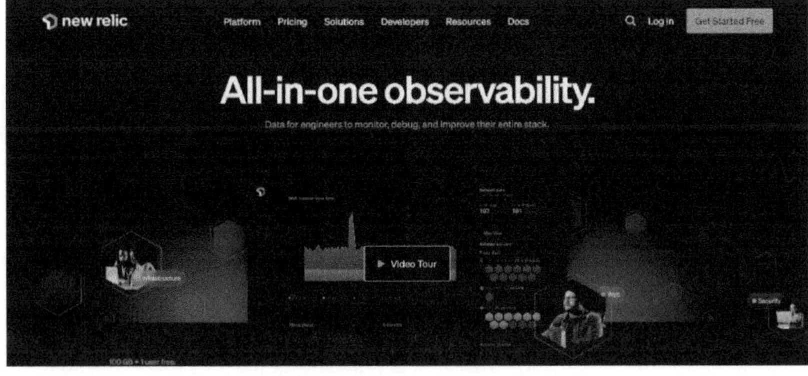

Source: - *(Relic, 2024)*

The goal of the cutting-edge full-stack observability platform New Relic is to enable engineers to efficiently plan, develop, deploy, and operate software. The platform creates a single source of truth for your whole system by offering a single interface for all telemetry data, including metrics, events, logs, and traces. In addition to gathering data, New Relic has strong analytic capabilities that facilitate prompt problem identification and speed up the troubleshooting process. It offers artificial intelligence support for more efficient issue solving and allows for smooth integration into current workflows.

4. Prometheus

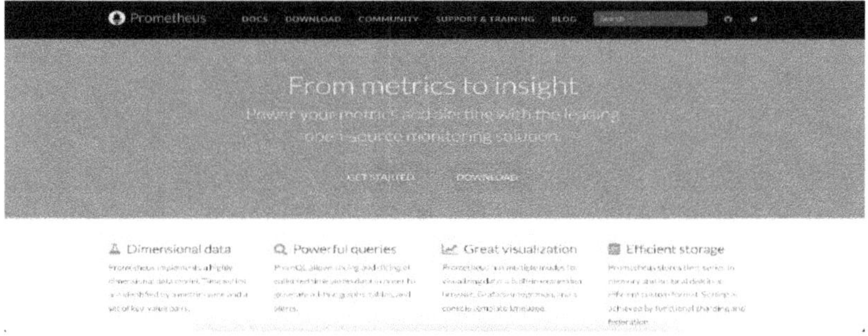

Source: - *(Prometheus, 2024)*

Prometheus is a top open-source monitoring system that improves your metrics and alerting operations with dimensional data modelling, robust query capabilities using PromQL, effective storage, and accurate alerting. It offers a complete solution for producing insights from metrics in an easy-to-deploy package because to its operational simplicity, interaction with many visualisation tools like Grafana, multiple client libraries for simple service instrumentation, and reliable alerts based on configurable PromQL.

5.   Dynatrace

Source: - *(Dynatrace, 2024)*

Dynatrace is an AI-powered analytics and automation platform that streamlines cloud complexity and enables safer, quicker innovation. With the platform's full-stack monitoring and intelligent and automated observability across cloud and hybrid environments, hosts, virtual machines, serverless, cloud services, containers, Kubernetes, networks, devices, logs, events, and more can be continuously discovered.

6.   PagerDuty

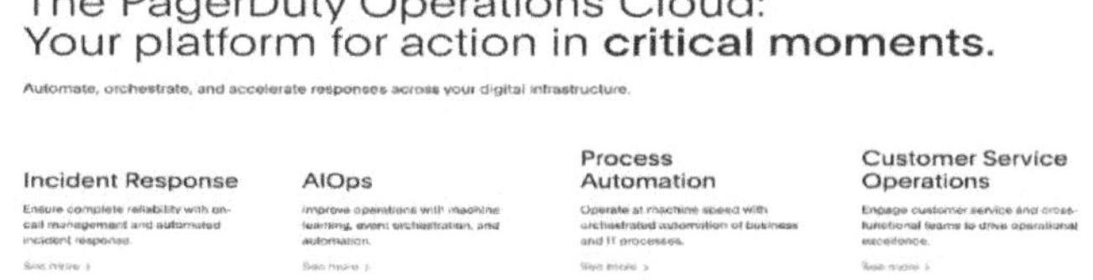

Source: - *(PagerDuty, 2024)*

A tool called PagerDuty is made to automate, coordinate, and speed up reactions throughout your digital infrastructure in times of need. In order to optimize operations and free up developers to concentrate more on their code, it provides capabilities like automated incident response, on-call management, process automation, machine learning for operations optimisation, and the opportunity to include customer care and cross-functional teams.

7.   Splunk

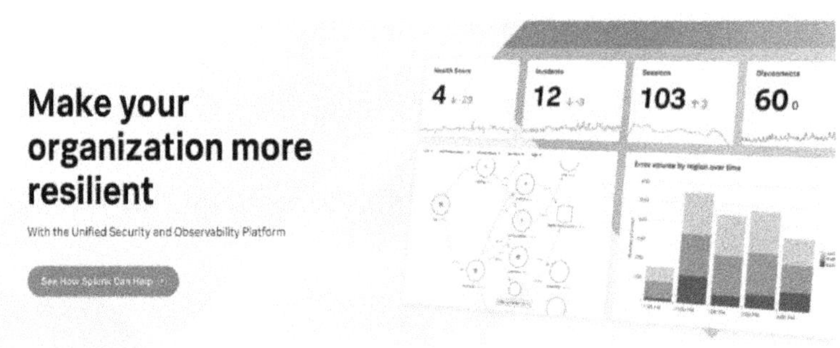

Source: - *(Splunk, 2024)*

Splunk is a single platform for security and observability that is intended to increase the security and resilience of digital systems. In order to assist security, IT, and DevOps teams in adapting, innovating, and providing for their clients, it provides features like actionable analytics for risk reduction, sophisticated threat detection for early incident prevention, and the ability to swiftly restore services during outages.

8.   Grafana

Source: - *(Labs, 2024)*

Grafana is a flexible platform for observability and visualisation that lets users query, visualise, and get alerts on data from a range of sources related to their business and technological operations. Grafana enables thorough monitoring across logs, metrics, applications, and infrastructure with capabilities like performance testing, multi-tenant log aggregation, high-scale distributed tracing, a scalable metrics backend, and a wide range of plugins.

9.   Elastic Stack

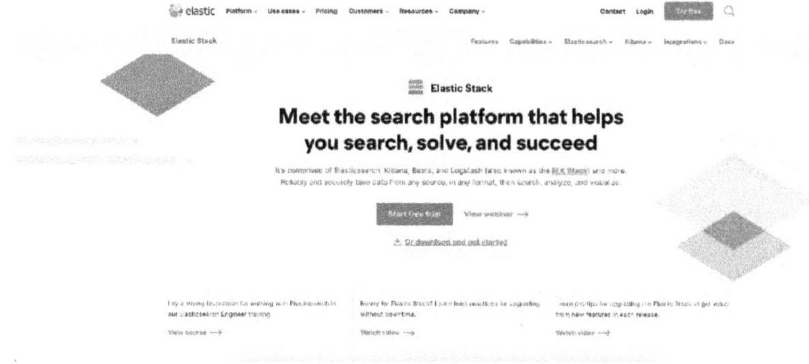

Source: - *(Elastic, 2024)*

Elastic Stack, which includes Elasticsearch, Kibana, Beats, and Logstash, is a robust search engine that can safely and consistently get data in any format from any source for analysis, visualisation, and hunting. With the option to install on-premises or on several cloud platforms, it provides quick, scalable data storage and search, real-time data analysis with Kibana visualizations, and connectors for ingesting data from many sources.

Cloud-based infrastructure management is essential to corporate operations. Businesses may obtain important insights into the functioning of their applications by utilising automated and manual cloud infrastructure monitoring solutions effectively. These technologies make it possible to record important performance indicators and carry out thorough analyses of company KPIs, which eventually improves operational effectiveness and decision-making.

Monitoring is about making sure your entire system is in line with your business objectives, not only about the state of operating systems or cloud resources. Maintaining peak performance and optimizing the advantages of your cloud trip may be made possible with the help of the appropriate monitoring tools.

AI performance management and optimisation solutions are critical to sustain high efficiency in the steady medium and long-term AI operations in the cloud frameworks. They allow for timely detection of performance problems due to constant monitoring of performance indicators such as response times, resource consumption, and error frequencies. While using AWS CloudWatch, Azure Monitor, and Google Cloud Operations Suite, companies can control performance notifications and ensure that the infrastructures are working at their most effective. They also help in capacity planning, because business needs to know how much of a particular resource they will need in the future, so that they do not over-buy or under-buy it. Finally, the combined approach of performance monitoring with optimization strategies affords reliable stability and performance, reduced operational costs, and the flexibility required in growing AI workloads.

5.5 Cost Management and Resource Allocation Strategies

Cloud computing provides businesses with unparalleled chances to increase the digital operations' cost-effectiveness, scalability, and agility. Only 30% of businesses, according to recent study, are fully aware of what they are paying for in the cloud, which makes it challenging to control expenses and guarantee operational effectiveness. Underutilized cloud resources, rapidly growing data, and hidden costs are just a few of the variables that make it difficult for many businesses to plan and manage their cloud budgets.

Cloud cost management (CCM) is a set of instruments and tactics used by businesses to cut costs and boost productivity when using the cloud. Before going over some cloud cost management best practices that may help you keep costs under control, this tutorial defines CCM and discusses why it's crucial for enterprises.

What is cloud cost management, and why is it essential?

Cloud cost management, also known as cloud cost optimisation, is the process of examining cloud services, apps, and infrastructure to make sure a business is only utilising the resources it truly requires. The flexibility to add new features to your cloud service at the touch of a button and auto-scaling can make recurring cloud charges unpredictable. The following are other elements that drive up cloud costs:

- **Under- or Mis-Utilized Cloud Resources.** More CPU power, storage space, or features than are required for a given task, and buying more for future instances instead of redistributing resources.

- **Poor data governance.** Paying to keep redundant or unnecessary data in cloud data lakes as you don't have the rules or tools to recognized, classify, compress, and delete cloud data in a sensible manner.

- **Hidden fees/complex billing.** Adding cloud functionality without fully understanding the expenses involved, having different payment plans for each cloud service or provider, and not being able to predict how consumption would change.

- **Shadow IT.** When customers or departments buy the cloud services they want without informing IT or determining if an already-existing solution might manage their workflow.

- **Surging data growth.** Monthly data storage costs are rising as a result of the constant generation and storage of more data in the cloud, particularly in data lakes and data mesh settings for business intelligence analytics and AI/ML training use cases.

In order to cut expenses and improve business support, CCM teams reallocate or remove resources as necessary. Furthermore, cloud cost management seeks to maximize cloud-based operations to boost workload efficiency, performance, and dependability—all of which help businesses get more out of their cloud expenditures.

6 Cloud Cost Management Best Practices

The following cloud cost management techniques might assist businesses in cutting down on waste and ongoing cloud expenses.

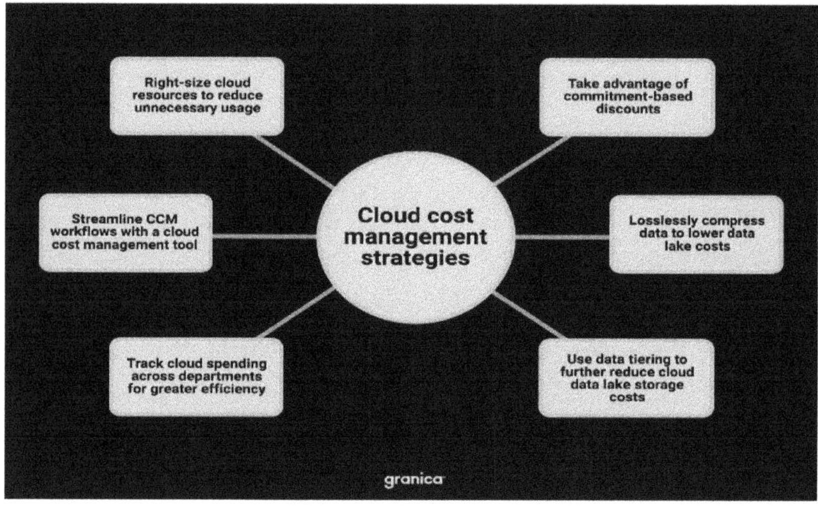

Source: - *(Granica, 2024)*

1. **Right-size your cloud resources to reduce usage**

    As one of the biggest cloud expenses is usually cloud computing resource utilisation (also known as CPU and GPU power), any inefficiencies might raise your ongoing cloud expenses considerably. For instance, IT departments frequently add more processing power to new cloud instances than are necessary for a particular job, which rapidly raises the hourly unit cost over what is truly needed. A lot of teams also fail to terminate existing instances or buy new ones instead of using the ones that are left unused, which might increase overall expenses.

In order to save ongoing expenses, right-sizing your cloud instances entails locating and removing utilisation inefficiencies so that CPU and GPU resources may be redistributed or terminated. IT personnel might manually monitor cloud usage, but such ineffective methods are prone to human error. With capabilities like resource identification and utilisation monitoring, using a cloud cost management application facilitates the automation of this process.

2. Take advantage of commitment-based discounts

Customers can reserve a specified capacity in a given area for a predetermined period of time (often one or three years) via Reserved Instances (RIs), Savings Plans, and other commitment-based discounts. The fact that these reductions provide substantial savings without altering your underlying infrastructure makes them an excellent top-line cost-cutting tactic.

On the other hand, long-term contracts include drawbacks, such as being tied into a pricing plan without the ability to grow or upgrade on demand, even if a supplier lowers rates. In order to enable other businesses to purchase RIs at an even greater discount, AWS users can sell their RIs on Amazon's Reserved Instance Marketplace.

3. Losslessly compress data for data lake cost reduction

Cloud data lake storage is another significant expense. Automatic scaling in conjunction with rapid data expansion can make recurrent cloud fees incredibly unpredictable. Finding high-priority data is quite difficult, particularly in situations with data lakes and data meshes. In order to maintain the quality of AI training or satisfy compliance standards, many teams simply lack the time and technological resources to precisely filter through large amounts of data. Frequently opting to be cautious, many businesses just retain everything and store that data in the highest costly storage class or tier.

Since Apache Parquet files are the foundation of cloud data lakes and lake houses, data compression reduces the physical size of data files without causing loss. This lowers the cost of cloud storage. As a result, data remains in its current tier but the effective unit cost of cloud data lake storage is reduced. Organisations can control expenses while maintaining data in the "standard" tier with full availability SLAs by using lossless data compression. Faster data transfer, reduced carbon footprint, and enhanced performance for I/O-bound applications are some other possible advantages of compression.

However, it requires time and money to compress and decompress data as it is read. Enterprise-grade lossless compression solutions designed for cloud data lake houses are essential to preserve scalability and even enhance data processing performance because the majority of off-the-shelf compression technologies aren't optimised for speed, cost, and scale in the cloud.

4. **Use data tiering to further reduce data lake storage costs**

Data compression by itself may not always be sufficient to achieve the required amount of cloud data lake cost reduction. Data tiering is the process of allocating data to the proper storage place based on its relevance and urgency. For instance, you may shift data for compliance or other recordkeeping needs to less expensive "cold" storage, but the data you're actively utilising for AI training and analytics should remain in costly "hot" storage.

Certain cloud providers, like AWS S3 Intelligent Tiering, provide automatic data tiering that transfers data according to the amount of time since the previous visit. But because "standard" data is more readily available than lower-tiered data, many organisations prefer to retain as much data in the standard tier as feasible. However, because conventional storage is more expensive, CCM teams should use compression and data tiering to assist reduce costs.

5.   Track cloud spending across departments

A lot of cloud services are so easy to use that departments may buy and implement them without help from IT professionals. The following dangers and inefficiencies are frequently caused by this situation, which is referred to as "shadow IT.":

1. Inadequate configuration or onboarding of cloud services prevents IT staff from patching vulnerabilities, keeping an eye out for any breaches, or deploying security technologies to safeguard them.

2. A department may decide to buy a new tool even if the business already pays for a cloud service that offers the same features.

3. It might be challenging for IT teams to develop precise budgets or put into practice efficient cost-management techniques if they don't fully comprehend how much the business actually spends on cloud infrastructure and services.

To get the most out of cloud cost management's efficiency and cost-saving advantages, you need a trustworthy way to monitor cloud expenditure and use across departments. Finding every application and service that a company uses is made easier with the use of cloud discovery tools. Certain cloud cost management tools further provide capabilities such as cost allocation tags, which enable IT to monitor cloud expenditures by workload, department, or other useful categories.

6.   Streamline CCM workflows with a cloud cost management tool

To help with the above-mentioned cloud cost management solutions, organisations can select from a range of software tools and platforms. Many businesses combine two or more technologies to provide all the features they need because no one platform delivers every component. The most widely used cloud cost management solutions are shown here, along with some of the capabilities they offer.

This makes cost control and resource provisioning crucial for creating the most cost-effective AI powered cloud systems. Cost control means tracking the use of resources, finding out unused resources and avoiding wasteful expenditure. AWS provides a service called AWS Cost Explorer, Azure provides Azure Cost Management, Google Cloud provides Google Cloud Billing, all of which provide insights into the cost of their usage and give organizations information on how they can reduce costs. Using reserved instances and having auto-scaling mechanisms and rightsizing can also help reduce costs further while maintaining performance.

Resource allocation strategies are aimed at the distribution of computing capabilities and storage as needed in context of various AI workloads. To make the right resource used for the right tasks, workload-specific resource pools like GPU instances to handle powerful AI model training and the lower-cost instances to handle lighter AI model inference. Thus, it is possible to balance performance and costs using demand-based policies, for example, demand-based provisioning or predictive scaling.

The challenge then becomes how to achieve cloud infrastructure that is balanced in terms of cost and resource utilization so that it can support large workload for AI while at the same time being cost effective. This is coupled by performance check, budgets alerts and capacity planning to ensure that resources are always in harmony with business demands. In conclusion, a proper systematic procedure improves both the financial profitability and efficiency of the systems implemented in the cloud AI hosting Platforms.

5.6 Real-World Examples of Scalable Cloud-AI Implementations

Real world case studies of cloud-AI at scale show how the underlying cloud technologies enable AI applications, enhances for massive workloads, scaleable, and can be fine-tuned for expansiveness, value, and robustness. With the increase in the complexity of the AI workloads organizations have started to leverage cloud infrastructure as the demand for elasticity and scalability continues to rise. These implementations show how cloud can be used to create AI pipelines that are manageable, and can handle workloads as they come, as a means of applying the cloud towards functional uses across industries and markets with respect to data handling and analysis, moderation, and more.

One of such examples is Strise.ai that offers services in artificial intelligence designed for data analysis. A key issue that Strise.ai had to deal with was the problem of handling lots of data and computation involved in the training of the various machine learning models. To this, the company resorted to Google Kubernetes Engine (GKE) responsible for managing containerized workload. Thanks to GKE, Strise.ai can adjust its infrastructure requirements without any problems, which is very advantageous in terms of resource allocation. During workloads that require a larger amount of processing power to process data, auto-scaling in GKE acquires new resources. In an ideal case the system scales down once the demand is low to avoid wastage of resources and high costs. This flexibility helps Strise.ai to sustain high performance and at the same time does not lead to over-provisioning of resources which also makes Strise.ai affordable and expansible. This approach is especially beneficial in the AI niche since resource usage can be highly fluctuant, and many tasks are computationally challenging.

In the same manner, Applica – an AI-based content moderation company – leverages AWS as the core technology covering its operations. The solution offered by Applica means that the content posted by the users can be filtered and moderated in real time to fit with the rules of the community. It also can be irregular because of the content that goes viral or receives more attention from users for some time. To deal with this, Applica utilizes the scalability of AWS since it simply means that the company can quickly increase or decrease the resources which it has hired depending on the current needs. With AWS Auto Scaling, Applica can be confident that its AI models will be up and running, and able to respond to demand without burdening its own systems. The flexibility of AWS infrastructure also supports cost cutting as it gets easily adjusted to consume less when the load is low, but still Applica can be sure they are paying for the needed amount of resources.

For instance, an international online store deploys Microsoft Azure to operate AI computations that include inventory and recommendation systems. By using Azure, the e-commerce company is easily able to scale the machine learning models and scale up or down the resources as and when required. Azure VMs and managed Kubernetes services can be used by the platform to scale up the computational resources required to process big data during the sales rush, such as back Friday or seasonal sales promotion. At these times, the company needs to make sure personalized recommendations and inventory forecast to execute well. The platform cuts down resources after the peak period, hence reducing costs. This is not only helps improving operational effectiveness, but also helps to ensure that services based on artificial intelligence always quickly scale and adapt to new conditions, increased traffic, and changing consumer behavior.

The above examples show how organisations can apply the Cloud computing technology to accommodate large scale AI. Google Cloud, AWS, and Azure are some of the cloud platforms that avail efficient tools to run AI operations for businesses. The resource on-demand characteristic of cloud computing facilitates the accomplishment of performance demands at high use times and saves costs at other times. Additionally, approaches like containers in production using kubernetes, the use of serverless computing, and auto-scaling that are native to cloud are gradually becoming fundamental parts of AI solutions, making corporations flexible and cheap while not diminishing the quality of their AI systems.

In sum-up, the case studies of simple and extensive AI application in the cloud environment hold good lessons for those organizations that are in the process of designing AI systems to have a high performance-to-efficiency ratio. In the future as AI progresses the value in the scale of the resources needed for its development will be one of the most critical aspects. By leveraging cloud technologies that underpin elasticity and automation of AI platforms, various companies will be better placed to cater for the future growth of AI applications hence be able to stay relevant in the current fast evolving market environment.

5.7 Chapter Summary

Chapter 5 provided an in-depth exploration of scalability and performance optimization in Cloud-AI systems, focusing on essential strategies and tools for enhancing the efficiency and flexibility of AI workloads in the cloud. It covered the design of scalable cloud architectures, highlighting the importance of creating adaptable infrastructures that can respond to dynamic AI demands. The chapter also emphasized the role of elasticity and load balancing in ensuring optimal performance during fluctuating workloads, as well as the use of performance monitoring and optimization tools to maintain system health and cost-efficiency. Additionally, it discussed cost management and resource allocation strategies, offering insights into balancing resource use with financial goals through smart scaling and provisioning. Real-world examples, such as those from Strise. ai and Applica, demonstrated the successful implementation of scalable AI systems using cloud platforms like GKE and AWS, showcasing the practical application of these concepts. Ultimately, the chapter highlighted how integrating these strategies can enable organizations to optimize their Cloud-AI systems for better performance, cost-effectiveness, and adaptability in a rapidly changing technological landscape.

Multiple Choice Questions (MCQs)

1. **What is the primary focus of designing scalable cloud architectures?**

    a. Minimizing software development time

    b. Ensuring the system can handle increased workloads

    c. Reducing hardware dependencies

    d. Improving user interface design

2. **Which term refers to the ability of a cloud system to automatically adjust resources based on demand?**

    a. Elasticity

    b. Scalability

    c. Redundancy

    d. Availability

3. **What is the primary role of load balancing in AI workflows?**

    a. Reducing the size of AI datasets

    b. Distributing workloads across multiple resources

    c. Simplifying cloud security protocols

    d. Optimizing AI model accuracy

4. **Which tool is commonly used for performance monitoring in cloud environments?**

    a. TensorFlow

    b. AWS CloudWatch

    c. Docker

    d. Kubernetes

5. **What is the main goal of cost management in cloud-AI systems?**

    a. Increasing the speed of AI model training

    b. Allocating resources efficiently to reduce expenses

    c. Automating all cloud workflows

    d. Enhancing cloud-native security

6. **What is a key advantage of scalable cloud-AI systems?**

    a. Fixed resource allocation

    b. Improved handling of variable workloads

    c. Reduced cloud platform features

    d. Enhanced data redundancy

7. **Which strategy is effective for optimizing resource allocation in cloud systems?**

    a. Over-provisioning resources

    b. Monitoring usage and scaling dynamically

    c. Avoiding performance monitoring tools

    d. Disabling elasticity features

8. **In real-world scalable cloud-AI implementations, which feature is critical for handling peak traffic periods?**

    a. Static resource allocation

    b. Reduced AI training times

    c. Elastic load balancing

    d. Simplified storage management

9. **What is the primary benefit of performance optimization tools in cloud-AI systems?**

    a. Simplifying the user interface

    b. Identifying and resolving bottlenecks in real time

    c. Enhancing the accuracy of AI predictions

    d. Reducing the need for cloud security

10. Which of the following best describes a scalable cloud-AI system?

    a. A system with fixed resource usage

    b. A system with limited support for AI workloads

    c. A system that focuses on manual performance tuning

    d. A system that adjusts capacity to meet demand

Answer

1	2	3	4	5	6	7	8	9	10
b	a	b	b	b	b	b	c	b	d

INDUSTRY APPLICATIONS OF CLOUD-AI INTEGRATION AND AUTOMATION

6.1 Chapter Overview

In chapter 6, the current and potential applications of Cloud-AI integration and automation in different industries are explained, as well as how organizations in different sectors are capitalizing on the possibilities offered by the integration of cloud computing and artificial intelligence technologies. This chapter provides a thorough examination of many industries, including healthcare, banking, manufacturing, retail, and logistics, using artificial intelligence (AI) technology combined with cloud infrastructure. It explores major advantage associated with this integration such as; better scalability, real time processing of data, cost reduction and handling of complicated processes. Drawing on cases and examples of Cloud AI initiatives and application, the chapter reveals how Cloud-AI solutions can help businesses to enhance performance, expand customer satisfaction and remain effective in the digital economy.

6.2 AI-Driven Cloud Solutions in Healthcare

In the healthcare industry, cloud computing streamlines and secures the exchange of medical records, automates backend processes, and even makes it easier to develop and manage telehealth applications. The healthcare sector is more efficient when using the cloud, and expenses are also reduced.

Healthcare spending on cloud computing is expected to reach a global market value of over $89 billion by 2027 due to its rapid growth. Furthermore, with a projected CAGR of 32% by 2027, Infrastructure as a Service (IaaS), a cloud computing paradigm that is popular for moving healthcare infrastructures to the cloud, is now the fastest growing cloud service (Builtin, 2024).

Benefits of Cloud Computing in Healthcare

- Accelerates clinical analyses and care processes.

- Automates data processing and scalability.

- Increases patient data accessibility.

- Reduces network equipment and staff costs.

- Reduces risk of data loss.

The cloud's many benefits and potential benefits are difficult to overlook, which is why adoption is growing even if there are still cloud sceptics, many of whom raise privacy concerns because they are unwilling to share extremely sensitive information with an outside provider.

As operations shift to cloud-based software and servers, hospitals and other healthcare organisations are seeing a decline in the use of internal IT professionals, similar to many other businesses. Supporters of the change claim that it will result in a number of benefits, such as reduced expenses and the capacity to expedite the examination of crucial elements of care, such clinical notes, which should ultimately result in improved treatment.

In addition to payroll being significantly reduced and data being handled much more quickly without huge in-house IT teams, there are no on-site servers to destroy or gravely compromise in the event of malevolent cyberattacks or natural catastrophes. Also crucial is automated scalability.

"It's much easier to increase the scale and capacity in a cloud-based environment if you have an application that's growing exponentially like EHRs (Electronic Health Records) tend to do," Bob Krohn, partner and healthcare practice lead at international research and advisory firm ISG, told Healthcare Dive.

Perhaps most importantly, the availability of comprehensive patient data from various sources to physicians, nurses, and other healthcare professionals removes the need for extensive networks and intricate security measures. Patients receive the necessary information, drugs, and treatments more rapidly and precisely as a result. That's the objective, anyhow.

The combination of cloud computing and healthcare has enormous potential to enhance a number of healthcare-related functions, including medication adherence, drug anti-theft and counterfeiting measures, telehealth and virtual care, resource inefficiency, personal data privacy, and the uniformity of medical records, according to data gathered by the healthcare division of Tokyo-based Renesas Electronics Corporation. Therefore, it's not surprising that more cloud providers are fighting for a share of the healthcare market.

The businesses listed below are assisting hospitals and other healthcare institutions in using the cloud to improve patient care and streamline operations.

1. Microsoft

Figure 6.1: Microsoft logo

Source: - *(Builtin, 2024)*

Microsoft is well-known for developing a wide range of software programs, many of which are used in the medical field. Azure is a cloud-based platform that may be used to track patient insights and analytics within the cloud when combined with AI and IoT technologies. Holographic pictures, including any accessible data stored in the cloud, can be projected onto the wearer by another Microsoft product, the HoloLens 2. This enables medical professionals to access patient information and teamwork tools while diagnosing, treating, and operating on patients.

2. Pfizer

Figure 6.2: Pfizer logo

Source: - *(Builtin, 2024)*

In addition to using cloud services for its projects since 2016, Pfizer is a pharmaceutical and biotechnology company that recently gained attention for its collaboration with BioNTech to develop a COVID-19 vaccine and its collaboration with Amazon Web Services to develop cloud-based solutions that are intended to expedite and enhance the development, manufacturing, and distribution processes for clinical trial testing.

Cloud Computing for Electronic Health Record Management

1. GHX

Figure 6.3: GHX logo

Source: - *(Builtin, 2024)*

GHX runs a global network of cloud-based supply chains that link hundreds of healthcare organisations. The SaaS company enhances visibility into the products used in patient care by healthcare organisations and streamlines supply chain procedures. Through API integrations with well-known industry solutions, its GHX Data Connect product seamlessly syncs with the content of electronic health records.

2. 1upHealth

Figure 6.4: 1upHealth logo

Source: - *(Builtin, 2024)*

A range of cloud-based solutions from 1upHealth are intended to address data-related issues in the healthcare sector. 1upHealth's technology is used by payers, providers, digital health firms, and life sciences organisations to automate processes, provide patients with convenient access to their records, compile clinical and claims data for analysis, exchange comprehensive population-level patient data, and effectively handle other crucial data-related tasks.

3. Tiger Connect

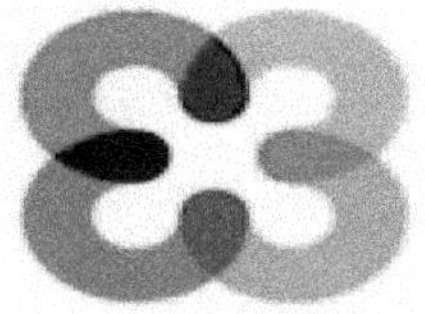

Figure 6.5: TigerConnect

Source: - *(Builtin, 2024)*

A cloud-based software firm called Tiger Connect gives medical professionals a messaging platform that complies with HIPAA regulations and enables them to exchange papers, patient data, records, and communications. It also makes it possible to work together on treatment plans and paperwork. One use for the program is physician on-call scheduling, which is made possible by a scheduling system that can almost immediately generate a year's worth of schedules.

Managing Patient Care, Not Patient Data

"Cloud adoption will grow at a very quick pace, "One reason is that major firms like Amazon, Google, and Microsoft realized very early that many hospitals will not be able to continue using on-premises data centres. Additionally, they are certifying their data centers for safe harbour compliance procedures and HIPAA (Health Insurance Portability and Accountability Act of 1996) compliance. Our work is made easier if we can obtain guarantees from our partners that they are high-tech and HIPAA-certified. It is not required to oversee infrastructure. Furthermore, we can manage it without putting in place a comprehensive security procedure. (Hashem et al., 2015)dedicated space, and software. Massive growth in the scale of data or big data generated through cloud computing has been observed. Addressing big data is a challenging and time-demanding task that requires a large computational infrastructure to ensure successful data processing and analysis. The rise of big data in cloud computing is reviewed in this study. The definition, characteristics, and classification of big data along with some discussions on cloud computing are introduced. The relationship between big data and cloud computing, big data storage systems, and Hadoop technology are also discussed. Furthermore, research challenges are investigated, with focus on scalability, availability, data integrity, data transformation, data quality, data heterogeneity, privacy, legal and regulatory issues, and governance. Lastly, open research issues that require substantial research efforts are summarized. © 2014 Elsevier Ltd.","author":[{"dropping-particle":"","family":"Hashem"," given":"Ibrahim Abaker Targio","non-dropping-particle":"","parse-na mes":false,"suffix":""},{"dropping-particle":"","family":"Yaqoob","given":"Ibrar","non-dropping-particle":"","parse-names":false,"suffix":""},{"dropping-particle":"","family":"Anuar","given":"Nor Badrul","non-dropping-particle":"","parse-names":false,"suffix":""},{"dropping-particle":"","family":"Mokhtar","given":"Salimah","non-dropping-particle":"","parse-names":false,"suffix":""},{"dropping-particle":"","family":"Gani","given":"Abdullah","non-dropping-particle":"","parse-names":false,"suffix":""},{"dropping-particle":"","family":"Ullah Khan","given":"Samee","non-dropping-particle":"","parse-names":false,"suffix":""}],"container-title":"Information Systems","id":"ITEM-1","issue":"July","issued":{"date-parts":[["2015"]]},"page":"98-115","title":"The rise of \"big data\" on cloud computing: Review and open research issues","type":"article-journal","volume":"47"},"uris":["http://www.mendeley.com/documents/?uuid=c5944c07-8e82-4388-9100-c545d52d17b4"]}],"mendeley":{"formattedCitation":"(Hashem et al., 2015.

In the healthcare sector, AI-based cloud solutions have also been crucial in improving information sharing and communication across healthcare institutions. Cloud platforms help to provide the opportunity for secure data sharing between hospitals, research centers, and diagnostical laboratories for improving patient treatment and boosting the discovery of medications. For example, cloud data lake with artificial intelligence can integrate the records of the patient from multiple sources which would result in better decision-making regarding patient's health. It equally enables the large-scale clinical trials whereby the substantial dataset may be managed and analysed, resulting in early research end-results and faster drug development (Bajwa et al., 2021).

In addition, AI in the cloud has a greatly enhanced role in population health and predictive analytics. It can be used to generate big data analytics in patient data to determine the trends prevalent in the public healthcare systems, forecast disease spreading seasons, and contain disease outbreaks and strategize for preventive care in advanced healthcare organizations using machine learning models located in cloud services. For instance, it is possible to predict flu seasons or chronic diseases trends, by analysing health data, then help the public health governments to make better decisions with the available resources. In the future, as the development of cloud infrastructure progresses, the integration of artificial intelligence and cloud technologies will intensify the generation of innovations, enhancement of the availability of healthcare services at reasonable prices and enhance the quality of the healthcare services globally.

6.3 Revolutionizing Financial Services with Cloud-AI Automation

Cloud-AI automation is the next generation solution taking over the financial services industry by rationalizing work processes, decision-making, and customer experiences. The use of artificial intelligence in combination with cloud technologies increases the effectiveness of routine job like fraud detection, risk evaluation, and compliance monitoring. The cloud computing helps the financial firms in getting large computational power required for developing AI models for processing the real-time data and get quicker insights on market trend and customer behaviour. It also enables tailored monetary services including artificial intelligence in managing individual or corporate wealth by offering proper funds investment recommendation depending on one's wealth status (Ionescu & Diaconita, 2023)they are increasingly harnessing advanced data management technologies such as artificial intelligence and cloud computing. This paper presents a comprehensive review of how these tools transform financial decision-making in various domains and applications. We analyzed both foundational and recent advancements using a rigorous methodology based on the PRISMA 2020 guideline. Our findings indicate that many major financial institutions are adopting AI-driven solutions to potentially enhance real-time risk assessment, transactional efficiency, and predictive analytics. While they bring benefits like faster decision-making and reduced operational costs, they also pose challenges like data security and integration complexities that require further research and development. Looking ahead, we envision a more integrated, responsive, and secure financial ecosystem that leverages the convergence of AI, cloud computing, and advanced data storage. This synthesis underscores the significance of contemporary data management solutions in shaping the future of data-driven financial services, offering a guideline for stakeholders in this evolving domain.","author":[{"dropping-particle":"","family":"Ion escu","given":"S. A.","non-dropping-particle":"","parse-names":false,"suffix":""},{"dropping-particle":"","family":"Diaconita","given":"V.","non-dropping-particle":"","parse-names":false,"suffix":""}],"container-title":"International Journal of Computers, Communications and Control","id":"ITEM-1","issued":{"date-parts":[["2023"]]},"title":"Transforming Financial Decision-Making: The Interplay of AI, Cloud Computing and Advanced Data Management Technologies","type":"article-journal"},"uris":["http://www.mendeley.com/documents/?uuid=c48e2afa-d1bf-427b-a26b-203ac84e6161"]}],"mendeley":{"formattedCitation":"(Ionescu & Diaconita, 2023.

In addition, cloud-AI automation enables financial institutions to address the problem of regulatory compliance by automating data checks, as well as reporting and the identification of outliers. By having machine learning algorithms in the cloud, the system can analyse all the transactions for the presence of suspicious activities and prevent financial crimes as well as conform to international regulations at all times. The flexibility, provided by cloud platforms, also contributes to a faster implementation of AI-based solutions, for instance, virtual assistants for customer service and credit scoring predictions. In the context of growing digitalization of the financial industry, cloud-AI automation has become a crucial enabler of processes and cost optimization as well as the creation of innovative services (Blessing, 2024).

Cloud-AI automation has brought a major positive change in the banking industry where efficiency has been improved and the issue of risk minimized. Large banks use cloud AI for fraud prevention, transactional monitoring, credit risk analysis and more accurately. They also offer improved customer relations through the use of chatbots and virtual personal assistants for financial advice. With the help of the elastic cloud computing architecture, clients can securely process large data and therefore offer timely decision-making solutions and efficient services for compliance reporting and cost optimization to perform.

Revolutionizing Banking with AI-Driven Cloud Solutions

The signals of a technological revolution are therefore already evident and becoming more so in the banking sector as two trends—cloud computing and artificial intelligence (AI)—combine to revolutionise how banks operate and interact with their customers (Blessing, 2024).

As the companies concerned face the pressure to maintain competitiveness and continue to operate to serve their clients in the best possible way, they start focused on these technologies that can help to improve their offer in terms of quality and performance.

This introduction therefore forms the background through which the potential and risks arising from the integration of AI driven cloud solutions in the banking sector will be discussed.

It will look at how these technologies have evolved into tools that reinvent efficiency and cut costs and above all how they help banks deliver differentiated and more secure products to the consumer, thereby helping to reshape the traditional doctrine of banking.

The Paradigm Shift: AI and Cloud Technology in Banking

The banking industry is amidst a major evolution that is being spurred by the use of emerging technologies such as AI and cloud. These are innovations that augment existing processes but are simultaneously creating innovative business models and services.

AI, together with the cloud, addresses the challenge of handling enormous streams of data and making instant decisions, helping the banks to optimize customer experience. Thus, this change is not merely about new gadgets, it is about the organization getting ready for a digital world.

They have now become capable, for instance, of providing real-time fraud detection and prevention, of accurately predicting customer behaviour, and of providing clients with highly individualized financial consultancy.

Enhancing Customer Experiences with AI-Driven Insights

Perhaps one of the biggest advantages of AI applications in banks is the use of data and analytics for every single client through customer relationship management, which is currently revolutionizing how banks engage with their customers.

With the help of AI algorithms, it would be possible to process an enormous amount of customer data and find patterns and trends that an unaided eye could hardly see. This kind of analysis helps the banks to gain higher levels of differentiation in order to better satisfy the needs and wants of customers or clients.

For instance, the use of AI is to hypothesise when a customer may be in need of a loan given their financial activity or rather financial history. AI can also recommend investments that meet the customer's investment profile and financial risk factors, which is something that had never been possible before.

The use of AI, chatbots, and virtual assistants in organizations' customer service delivery is a major boost, as it is available all the time. These smart others can answer virtually all questions from as basic as presenting account balances or transaction records to as complex as financial consulting or loan documentation.

The time saved by the efficient handling of common queries, likely by using AI solutions to provide the answers, lets constitutive human agents concentrate on more complex and less repetitive questions that are likely to require more individual handling, leading to better overall service delivery and customer satisfaction.

AI integration into customer service is also not uncomfortable as it is very effective in making the customer service very efficient in a way that customer needs various levels of assistance at various times and their needs are promptly and adequately addressed appropriately.

Cloud-AI automation is a continuing force of disruption throughout the financial services segment, although its influence is perhaps most obvious in banking. Real-time anomalous analysis of the data that is hosted in cloud applications can be as included AI tools in the detection of fraud before the occurrence of the actual event while the tools provide solutions for regulatory compliance, through validation and an automated report generation in the data. Also, it enables the fast expansion of the application of AI solutions in banks, ranging from unique financial offerings to algorithmic trading and risk management. Cloud-AI technologies will continue to gain popularity which can help to improve the efficiency of financial institutions, decrease their expenditures, and enhance the performance of data-driven services in response to the emerging trends on the market.

6.4 Smart Cities: IoT, AI, and the Cloud Convergence

Smart cities are the integration of IoT, AI and cloud computing wherein the development and scope of smart cities are transforming the efficiency and effectiveness of the different cities around the globe. Sensors, cameras, and smart meters are among the IoT devices that constantly collect plenty of real-time data from urban settings, traffic flow, energy use, and air quality, among others. Cloud platforms supply the fundamental architecture for managing this kind of data and AI algorithms process the data and generate insights and recommendations for city planners and relevant authorities. This kind of data management approach assists in making core service areas, such as waste disposal, traffic signalization, and electrical supply more sustainable and integrated, thus making urban living more efficient and functional (Whaiduzzaman et al., 2022).

Automated real-time prediction is an important aspect of smart cities' focus on enhancing mobility and minimizing traffic in metropolitan areas. Taxi, buses, and even pedestrian paths that are IoT connected can help train AI models to predict traffic congestion and change signals accordingly within real time. Cloud technology also supports dynamic route planning of public transportation networks in real-time dependents on commuter traffic patterns. Also, smart parking systems employ AI and IoT to assist drivers to parking spaces hence saving on fuel and causing less pollution. These advancements help create less environmentally intrusive and more commuter friendly urban environments (Elassy et al., 2024).

In smart cities, IoT, AI and cloud technologies are being used to improve the area of public safety and security. Some of the applications of AI include monitoring video feeds of connected cameras to capture any unusual activities or intrusion to alert the policing authorities. Cloud also helps make data available in real-time across departments so that, during emergencies, there is faster coordination. In addition, use of intelligent systems in disaster response can reduce the effects of natural disasters through analysis of climate data, earth quakes, floods etc to enable cities prepare well.

The last critical area that benefits from this convergence is an environmental sustainability area. Smart cities use data on cloud and artificial intelligence to control carbon footprints through energy management. Iot sensors when integrated with smart grids and co-ordinated by the powers of AI can help in the distribution of electricity more efficiently by adapting to demands they notice. Furthermore, AI can predict the energy requirements and enable the connection of RE sources such as solar and wind energy. As a result, the pro-environmental orientation of smart city development guarantees that such cities remain resource-friendly when catering to the increasing needs of people living in cities (Bibri et al., 2024)in turn, impacted smart eco-cities, catalyzing ongoing improvements and driving solutions to address complex environmental challenges. This aligns with the visionary concept of smarter eco-cities, an emerging paradigm of urbanism characterized by the seamless integration of advanced technologies and environmental strategies. However, there remains a significant gap in thoroughly understanding this new paradigm and the intricate spectrum of its multifaceted underlying dimensions. To bridge this gap, this study provides a comprehensive systematic review of the burgeoning landscape of smarter eco-cities and their leading-edge AI and AIoT solutions for environmental sustainability. To ensure thoroughness, the study employs a unified evidence synthesis framework integrating aggregative, configurative, and narrative synthesis approaches. At the core of this study lie these subsequent research inquiries: What are the foundational underpinnings of emerging smarter eco-cities, and how do they intricately interrelate, particularly urbanism paradigms, environmental solutions, and data-driven technologies? What are the key drivers and enablers propelling the materialization of smarter eco-cities? What are the primary AI and AIoT solutions that can be harnessed in the development of smarter eco-cities? In what ways do AI and AIoT technologies contribute to fostering environmental sustainability practices, and what potential benefits and opportunities do they offer for smarter eco-cities? What challenges and barriers arise in the implementation of AI and AIoT solutions for the development of smarter eco-cities? The findings significantly deepen and broaden our understanding of both the significant potential of AI and AIoT technologies to enhance sustainable urban development practices, as well as the formidable nature of the challenges they pose. Beyond theoretical enrichment, these findings offer invaluable insights and new perspectives poised to empower policymakers, practitioners, and researchers to advance the integration of eco-urbanism and AI- and AIoT-driven urbanism. Through an insightful exploration of the contemporary urban landscape and the identification of successfully applied…","author":[{"dropping-particle":"","family":"Bibri","give n":"Simon Elias","non-dropping-particle":"","parse-names":false,"suffix":""},{"dropping-particle":"","family":"Krogstie","given":"John","non-dropping-particle":"","parse-names":false,"suffix":""},{"dropping-particle":"","family":"Kaboli","given":"Amin","non-dropping-particle":"","parse-names":false,"suffix":""},{"dropping-particle":"","family":"Alahi","given":"Alexandre","non-dropping-particle":"","parse-names":false,"suffix":""}],"container-title":"Environmental Science and Ecotechnology","id":"ITEM-1","issued":{"date-parts":[["2024"]]},"title":"Smarter eco-cities and their leading-edge artificial intelligence of things solutions for environmental sustainability: A comprehensive systematic review","type":"article"},"uris":["http://www.mendeley.com/documents/?uuid=1a225779-23eb-4c49-8cf4-9bdd8f75aa47"]}],"mendeley":{"formattedCitation":"(Bibri et al., 2024.

Finally, IoT, AI, and cloud computing offer a premise for utilizing data in managing city development and creating better administration, more efficient regulation, and better quality of life for citizens. This, as smart cities progress further, will contribute to the core of the technological integration in smart cities to promote development of efficient and sustainable smart cities throughout the course of rapid urbanization.

6.5 E-commerce and Retail: Personalization Through Cloud AI

AI's integration into cloud e-commerce and retail has transformed customer experience personalization in an e-commerce environment and boost satisfaction. Cloud platforms containing AI can compute large volumes of consumer data including the browsing patterns, buying habits, and demographic details instantly. This data is then used to accomplish product recommendations, targeted marketing, and variable pricing. For instance, the Amazon uses the cloud Artificial Intelligence to understand the customer's behaviour and to recommend specific products, which enhances the chances of purchase and loyalty (Patil, 2024).

The best example of cloud AI in the retail business is product recommendations based on machine learning capabilities. This can be done over and over with the customer interactions and the feedback given by the customers, thus making the suggestions produced by the AI systems more appropriate to the customer's needs. Cloud platforms enable the availability of the large-scale capacity necessary to implement millions of customer interactions at a time so that recommendations are always timely and accurate. It is a win-win situation for both the retailer and the customer since the shopping experience is enhanced as well as more sales for the retailer are made due to offering consumer appealing products (Haleem et al., 2022).

In addition, the examples of recommendation services, cloud AI allows implementing dynamic pricing for the products which may be adapted according to the demands, customer behaviour, and competitors' prices. AI models deployed on various cloud solutions can also sort through daily sales, weekly, monthly sales and other dynamics to set optimal prices for product instantly. This information can then be used by the retailers to provide and promote price promotions or bundled prices and still make good profits. For example, by applying cloud AI, during festive seasons, it is possible to offer discounts for wholesale customers and at the same time be viable for the retailer.

Another area that also seen a new trend in usage of cloud AI based virtual assistants and chatbots are e-commerce and retail. These include applications that use the cloud, there are those customer support applications that are opens 24/7 for customer seeking general information such as shipment, returns and any other information about the product. Even support issues which are complex can be solved using present day AI methodologies through considering the customers feeling and past experiences and coming up with a response. Such automation entails lower work pressure on human support teams in regard to these aspects together with reliability and efficiency to customer satisfaction and loyalty.

Other applications of cloud-based artificial intelligence include inventory control as well as demand forecasting. AI systems assist firms to predict prospective demand alterations depending on past sales as well as current trends; a trader does not suffer from stockout or overstocking. It is certain that through the utilization of cloud platforms, such insights can be availed at different store locations on how best to allocate resources and contain the costs of operation. For instance, AI can be implemented in that it estimates the consumption rate of retailers on particular products during promotion or new product release to enable them to meet consumer expectations.

In summary, Cloud AI will help e-commerce and retail industries to offer consumers and retailers a better experience, efficiency and orientation in their services. Cloud AI also extends to stay relevant in the constantly evolving market of business with the help of services such as recommendation tools, dynamic pricing, AI-based customer support, demand forecasting, etc. It is clear that the cloud AI will continue to develop to better assist retailers to meet their customer needs hence, creating long term customer relations.

6.6 Manufacturing: Cloud and AI in Industry 4.0

The combination of cloud computing and artificial intelligence (AI) in the manufacturing environment is viewed as being one of the primary hallmarks of Industry 4.0 in which conventional manufacturing process become smart process. IaaS encompasses the capabilities of cloud platforms to establish data storage and analysis of the enormous amount of data produced in smart factories, while AI provides algorithms to analyze these data, maximize production performance, minimize downtime, and support operational decision-making. The real-time data analysis application provides for the manufacture the means of higher speeds of production, enhancing quality of production and most essentially reducing the costs by enhancing a competitive environment in manufacture (Zhong et al., 2017).

AI-based cloud computing enables one of the most impactful changes in Industry 4.0 – predictive maintenance. IoT sensors are generally installed on the machinery where they are able to constantly sense different attributes like temperature, vibration levels and speed of operation. This information is then sent to cloud solutions where analysing patterns and identifying future equipment failures happens. Thus, manufacturers are better placed to plan their maintenance activities and avoid unwanted and costly downtimes while at the same time maximizing the useful life of vital assets. This approach reduces many disruptions and also makes the production process more efficient throughout the chain (Rajak et al., 2023).

Manufacturing quality control and defects identification are also transforming by the implementation of AI. Through Computer Vision and Machine Learning models deployed via cloud, manufacturers can physically inspect products at the production line. It can detect any imperfections that may be hidden, or as small as surface flaws, or wrong assembly, with more precision than human eye and hand. Cloud AI supports precise and accurate data transmission and processing and hence allows manufacturers to promptly modify the assembly line to avoid poor quality production and excessive material loss.

Application of both cloud and AI in manufacturing has also boosted the aspect of supply chain optimization. The data stored on cloud base AI systems can help to evaluate the historical sales data, the performance of the suppliers and the trends in the market to make a better projection of the demand for the products. This predictive capability enables the manufacturers to control inventory, minimize lead time, and avoid stock out or overstocking. Also, AI can detect problems in a supply chain and recommend new sources of supply or different delivery pathways, which enhances general supply chain reliability and minimizes operational threats.

Moreover, AI or cloud technology is at the centre of the manufacturing process automation and robotics. Smart manufacturing utilises artificial intelligence to operate robotic arms, Computer numerical control machines and conveyance systems with very little or no human operation. The interconnected systems in distributed cloud platforms allow real-time sharing of data within the cloud and remote control. For instance, AI can improve efficiency of robotic systems as they work together to increase production rates without compromising on speed and accuracy.

This level of automation leads to increased effectiveness of the utilization of resources in the contemporary industries (Gao et al., 2024).

Hence, cloud and AI in Industry 4.0 have transformed manufacturing by introducing predictive maintenance, improved quality assurance, improved supply chain management and automation. Through manufacturers' increasing adoption of smart technologies, cloud and AI will remain key drivers of future growth and innovation of manufacturing industries, along with competitiveness at the global level. Smart manufacturing as the fundamental solution for sustainable industrial development: The ability to make informed decisions, optimize resource use and maintain the quality of final products is among the priorities of the industry's development in the digital age.

6.7 Education: Enhancing Learning Through Cloud-AI Tools

Cloud computing and AI are two of the key technologies in education that is transforming how knowledge is delivered, accessed and individualized for the learner and the tutor. AI enables educational institutions to provide personalized, real-time, data-intense learning environments, which help in making it easier to collaborate, access and enhance learning in educational institutions. Through the use of cloud infrastructure, large quantities of materials, student information, and achievement records can be stored, analysed, and possibly distributed in several applications. However, these data are analysed by AI algorithms to filter the content, create material for students, and improve the educational decision-making process (Parker, 2023).

One of the most obvious benefits of cloud-AI tools in education is that these technologies enable learners' customisation. The designed AI platforms can also use data such as assessment records, participation levels, and difficulty levels to design and recommend an individual learning path for students. Smart Sparrow and Dream Box Learning are adaptive learning platforms in which the machine learning models determine the difficulty and speed of the lessons depending on the progress of the students. Such platforms offer the relevant exercises, quizzes and materials for the students that may study in accordance with their needs and elect the proper level of difficulty, so these platforms minimize the gaps in knowledge and provide a personal approach to learning. Delivery of these tools is made possible through cloud infrastructure hence enhancing flexibility especially in the current remote learning environment (Nur Fitria, 2021).

Another revolutionary shift is AI-based assessment and grading software. Online-based platforms such as Grade scope and Turnitin use NLP as well as other machine learning approaches to mark essays, quizzes, and assignments more efficiently than does manual labour. Not only do they help educators avoid investing hours in writing comments, but they also guarantee student feedback uniformity. Taking assessment information to the cloud platform also enables teachers to store and share the data for long-term analysis in the hope of recognizing trends that may warrant attention.

In addition to assessments, cloud-AI tools have functions in handling administrative jobs and classroom administration. Teachers can adopt use of Google Classroom, Microsoft Teams, Schoology in order to handle lesson plans, roll calls, and announcements effectively. AI solutions are employed to help with filing resources, class timetables, and even dropping-out rates evaluations. This automation enables the teachers to pay much attention to factors such as instructional approach instead of spending most of their time on protracted paperwork.

One more effective area of cloud-AI tools application is virtual tutors and chatbots usage. The use of AI in the form of chatbots like IBM Watson Tutor and Brainly's AI help students get answers to questions regarding their course work, lecturers, and coursework, offer explanations as well as guiding a student through the course of solving a problem round the clock. These virtual assistants improve remote learning since they can provide responses independently of a human being, meaning the students can get help at any time, including nights. In face-to-face and blended as well as fully online learning contexts, these tools mediate the transition from traditional to online learning.

The use of technological applications that support collaborative works fosters learning experiences, more so by allowing for live interactions between students and instructors in educational clouds. There are many applications like Google Drive, Zoom, and Microsoft OneNote where the students can collaborate their documents, group projects, and video conferences even in different classroom settings. These Cloud AI tools can transcribe lectures, generate summaries, and provide real-time translation, so children from different language environments have the opportunity to enter international educational initiatives. This inclusivity is important in promoting education to reach the unserved populations of people.

Last but not the least, it is critically important in professional development and institutions' decision-making. AI integrated data analytical tools can be used by educational administrators for analysing institution's performance, analysing enrolment patterns, evaluating teaching strategies and methods. Cloud platforms allow for the collection and storage of massive amounts of data from several campuses or districts effectively to inform decision-making processes that lead to better design of curriculum, allocation of resources and overall strategic planning in education (Igbokwe, 2023).

Thus, cloud-AI tools are emerging as the key enablers of change in education by providing learners with customized learning experiences, automating evaluation, improving communication, and facilitating institutional administration. Such technologies enhance group and individual learning experiences of students and teachers while at the same time, eliminating bureaucracy. Indeed, the use of cloud-AI tools will remain on the rise, ensuring that quality education is accessible to every learner as well as ensure they are well equipped to handle the demands of a digital economy.

6.8 Media and Entertainment: Cloud AI for Content Creation and Delivery

Cloud computing and artificial intelligence have revolutionized the media and entertainment industry in a big way. These technologies have changed the way content is created, distributed and consumed because of a more efficient, personalized approach and analysis of data. Cloud solutions provide media companies with the means to store, process and distribute large amounts of media content in real-time and efficiently, while AI technologies bring value and convenience in the form of advanced content selection, processing and recommendation, which ultimately improves both creating content and end-user experience. In addition to rationalizing all processes, cloud computing and the use of AI technologies have created new opportunities for the creative side, communication with users, and business models (Prasad & Makesh, 2024).

AI is also being applied in content creation where it is used in the generation of concepts for scripts and music, creation of complete sequences for videos. For instance, AI can be used to create voices that can be used for voice over jobs, generate music from learned algorithms or give an initial edit of the video, where AI can find the important scenes, transitions, and audio tracks on its own. These artificial intelligence features, backed up with the cloud services, help content producers to cut working time drastically, enhance the stability and quality of the content, and dedicate more efforts to the creative process. AI technologies can also be used to forecast the preferences of the audience and as a result adjust putting out content to individual viewers' preferences to increase their engagement as well as the odds of share-ability (Anantrasirichai & Bull, 2022).

Another area of importance is the use of cloud-based AI tools in the distribution of content. In more modern platforms, especially in video streaming platforms like Netflix, Hulu, and in music streaming platforms like Spotify, AI is used to filter the recommendations based in the patterns of the users' preferences and behaviour. Through the pattern analysis of user's TV viewing habits, choices and feedback, AI, recommends content that is likely to be more enjoyable to the user hence enhancing viewership. These services are supported by the cloud infrastructure whereby customers are able to access large databases of content seamlessly regardless of high traffic at particular times. This scalability is crucial for platforms which require millions of people to interact and provide seamless, continuous experience (Sharma et al., 2024).

In the same way, analytics through artificial intelligence are applied in marketing information in the media and entertainment industries. Audience analytics is one way through which AI assists particular corporations to forecast trends, monitor content, and create ads. Recommendation systems can also forecast which content can excel in certain markets or demographics so that media firms can adjust their promotion methodologies. In addition, the AI tools can work in parallel with monitoring social media and online discussions to identify the audience's sentiments and engagement levels more instantly and more accurately.

In addition, cloud AI has enabled other improvements on media security to protect creators, distributors, and consumers from piracy and unauthorized access. The functions of the cloud-based AI empowered encryption, watermarking and DRM helps the organization to ensure that the stored media assets are not shared or duplicated without the consent and authorization of the organization. Such technologies assist in the protection of intellectual property and allowing the right audience to access the content at the right price.

In conclusion, cloud computing and AI have presented the media and entertainment industries with significant changes that have led to improved speed and efficiency of content production, improved customer engagement, and better levels of security protection. Specifically, in production, distribution, marketing and supply of content, these technologies have enhanced the efficiency of content creators, distributors and streaming services while providing consumers with better value for their money and a much better experience. With the development of AI and cloud technologies, they are going to occupy even larger positions in media and entertainment industries in the future as a tool for innovation and as a response to the constantly increasing demand for content.

6.9 Transportation and Logistics: Autonomous Systems and Cloud Integration

The application of auto system integration and cloud computation in the movement and distribution industries is changing the business rapidly. These are innovations that are revolutionising transport of goods and other resources as well as people across the world through better ways and time, space and resource efficiency, safety and sustainability. Automated systems like self-driving cars, drones, as well as robotic systems are continuing to appear in operating environments in order to lower the impact of human intervention, avoid mistakes, and increase the rate of delivery. Cloud computing is instrumental as it supplies the platform that is required in the management of large volumes of data from these systems to facilitate real time communication, monitoring and optimization and the logistics chain (Golightly et al., 2022).

In the transportation industry, self-driving cars, commonly known as AVs are transforming delivery services as well as passenger services. Such cars are fully dependent on the AI in the cloud for direction, control, and even the addition of safety measures. Self-driving trucks for instance are capable of driving on long hauls without drivers and can use real time traffic, weather, and road conditions in an effort to achieve maximum fuel utilization. The cloud infrastructure helps the continuous flow of data from the AVs where the fleet operators can monitor the performance of the vehicle, track shipments and make changes where necessary to ensure timely delivery. Predicting other possible problems that may occur and the remaining useful life of the vehicle components are also done by artificial intelligence and machine learning algorithms from the data collected by the sensors of vehicles, increasing the reliability of vehicles and decreasing the time of vehicle's no availability (Torres & George, 2023).

Cloud computing and autonomous technologies are being integrated in logistics to improve last-mile delivery and supply chain management. Small parcel delivery is another application area where fully autonomous drones and robots are being utilized to perform operations in privileged solutions faster, especially in the urban environment. These autonomous systems can use cloud platforms for planning routes, real time monitoring of inventory, and sending data to centralized systems for the latest status of orders. When implemented in warehouse management, companies in the logistics industry can benefit from sorting, packing, and inventory control since they become automated with minimal employment of people. This automation results in fewer errors, optimal resource deployment and improved overall productivity in the network of supply chain (Sornprom, 2024).

Cloud integration also has a great influence on the visibility and integration of logistic operations. It also allows the operators to monitor the position of the shipments, autonomous vehicles or drones for the shipment of products electronically, and also to get information instantaneously on any delay or disruption. The algorithm can identify areas of risks or congestion, and then avoid them in order to minimise delays. Source data of traffic reports, weather, and inventory information all gathered from the cloud provide better decision making leading to improved efficiency and lower operation costs.

Also, cloud computing improves the interaction between various players in the transportation and logistics industry. While shippers, carriers, warehouse managers and even customers can all access shared data over the cloud, this makes communication as well as the flow of information in the supply chain very efficient. Such a high level of integration helps to increase customer satisfaction due to prompt tracking and delivery information, credibility in logistics services.

Last, the application of autonomous systems and cloud computing in transportation and logistics are supportive of sustainability objectives. Self-driving cars, smart pathways, and artificial intelligence in delivery decrease the usage of fuel, decrease pollution, and decrease unnecessary spending. Through traffic analysis and optimal delivery planning, cloud-based AI systems of logistics companies save fuel and minimize environmental impact of transportation services. Thirdly, the mini logistics delivery system also decreases the environmental footprint by cutting the need for delivery trucks in some instances due to the use of drones (Mancino, 2022).

In conclusion, the combination of autonomous systems and cloud systems has brought a new generation of change in the transportation and logistics industries. These technologies are making the operations smarter, faster and more cost effective through automation of several key processes, improved routing and real-time data for decision making. Even as the scale of autonomous systems and cloud platforms increases these will become ever more important in the development of the international logistics and transportation industry, and the creation of more innovative, and sustainable, environment.

6.10 Chapter Summary

Chapter 6 discusses how Cloud-AI integration and automation change industries and how these technologies enrich organizational operations, customer experiences, and innovations. In the health care setting, the cloud computing and AI enhance patient care, reduce time and optimize data utilization in their treatment. For financial services, Cloud-AI automation delivers better decision-making, personalized products, and services, and fraud identification. IoT, AI and Cloud platforms help smart cities to improve the management of utilities and services while Cloud-AI assists the e-commerce and retail industries in customizing customer engagements and inventory management. Industry 4.0 as a manufacturing industry is transformed by the AI automation of manufacturing processes, maintenance, and supply chain. Education industry leverages Cloud-AI to enhance learning and teaching processes, as well as organizational administration. AI is widely used in media and entertainment industries for content development or dissemination, improving audience interaction and preventing piracy. Last but not least, autonomous systems integrated with cloud technologies serve transportation and logistics to increase delivery efficiency and decrease costs while being sustainable. In general, Cloud-AI integration is proving to be a disruptive force across these industries, providing affordable, accessible solutions that help organizations grow, deliver better services and become more sustainable in the digital economy.

Multiple Choice Questions (MCQs)

1. **How are AI-driven cloud solutions transforming healthcare?**

    a. By automating billing processes exclusively

    b. By replacing medical professionals with AI systems

    c. By eliminating the need for patient data storage

    d. Through advanced diagnostics and personalized treatments

2. **What is a key benefit of cloud-AI automation in financial services?**

    a. Faster physical branch expansions

    b. Manual handling of large datasets

    c. Enhanced fraud detection and risk management

    d. Reduction of customer personalization features

3. **What technology convergence is critical for building smart cities?**

    a. Blockchain and traditional IT systems

    b. IoT, AI, and the cloud

    c. Legacy systems and manual processes

    d. Virtual reality and standalone servers

4. **In e-commerce, how does cloud-AI enable personalization?**

   a. By limiting product recommendations

   b. Through customer behaviour analysis and tailored experiences

   c. By automating delivery logistics

   d. By standardizing product pricing

5. **Which of the following best describes the role of cloud and AI in Industry 4.0 manufacturing?**

   a. Automation and predictive maintenance of machinery

   b. Manual quality control processes

   c. Limiting scalability of production lines

   d. Decentralized and isolated operations

6. **How does cloud-AI enhance education?**

   a. By replacing teachers with AI systems

   b. Through adaptive learning platforms and personalized content delivery

   c. By reducing access to digital resources

   d. By limiting collaboration among students

7. **What is a major application of cloud-AI in media and entertainment?**

   a. Automating all content approvals

   b. Standardizing global content without localization

   c. Reducing media consumption options

   d. Enhancing content creation and personalized content delivery

8. **How is cloud integration enabling autonomous systems in transportation and logistics?**

    a. By eliminating the need for real-time data analysis

    b. Through route optimization and vehicle automation

    c. By limiting connectivity between vehicles

    d. By standardizing manual delivery methods

9. **What is the primary role of cloud-AI in smart cities?**

    a. Centralizing all data storage without analytics

    b. Replacing traditional utilities with AI-based services

    c. Reducing data sharing across city departments

    d. Enhancing real-time decision-making and infrastructure management

10. **In what way does cloud-AI benefit the e-commerce and retail sector?**

    a. By standardizing product categories globally

    b. By reducing the need for inventory management

    c. By enabling hyper-personalized marketing strategies

    d. By limiting payment processing options

Answer

1	2	3	4	5	6	7	8	9	10
d	c	b	b	a	b	d	b	d	c

Chapter 07

FUTURE TRENDS AND ETHICAL CONSIDERATIONS IN CLOUD-AI TECHNOLOGIES

7.1 Chapter Overview

Toward that end, this chapter is centred on future development, current trends, and the ethical setting of Cloud-AI solutions. In the current world, organizations are searching for a means to integrate with the use of artificial intelligence, and it is also a popular choice that people look for cloud solutions, making it imperative that one discover the importance and the future that AI cloud solutions has. Some of the subject areas are encompassing: the critical infrastructure systems, distributed systems, sustainability, and rights as well as progressive technologies.

7.2 Building Resilient and Scalable Cloud-AI Infrastructures

In the context of digitalization and modernization of enterprises' activities, companies are playing a significant role in constructing robust and reliable systems which can deal with varying demands and provide high availability and constant value for users. Modern trends such as Cloud computing, AI, DevOps and DataOps are stirring the way enterprises design and implement their systems. Such integration leads to the development of systems that are flexible and intelligent as well as scalable, enabling businesses to achieve operational efficiencies and even breakthrough. That is why cloud computing provides enterprises with the capabilities to scale and manage resources more effectively to satisfy customer demand during critical loading periods without significant amounts of money invested in IT infrastructure. It forms a basis from which complex and even more adaptable systems can be developed with a view that can be implemented to business change. IaaS provides high availability, fault tolerance, and cost-efficient solution, which is valuable in today's enterprise architecture enactment.

The most recent and prevalent area that has shaped system processes throughout the enterprise is Artificial Intelligence (AI). AI helps make better decisions – it analyses large sets of data and presents conclusions in several forms; the use of AI means specific processes can be automated; and AI can also help identify problems before they appear in the business environment. The use of AI in DevOps helps in the testing, deploying as well and monitoring of applications and software, hence enhancing the speed of its release. The use of analytical data predictive by the AI also improves the operation of the system by predicting additional capacity, bottleneck conditions and security breaches that need to be addressed in order to improve the effectiveness of the system as a whole.

Cloud computing can be viewed as the key element of the contemporary enterprise architecture since it meets business requirements in terms of scalability, flexibility, and, consequently, adaptability to the modern conditions of the market environment. Cloud-platform-based systems mean that an organization is no longer limited to having to build large-scale systems due to the increased workloads, traffic, or simply because the company wants to expand, all without the need to often invest huge amounts of capital. To this, the cloud has high availability as well as has the fault tolerance that is required for enterprise systems to remain.

Cloud Computing: The Foundation of Scalable and Resilient Systems

Available even when the hardware on which this system runs fails or when the networks connecting this system to other systems fail. There are different advantages of cloud computing among which the fast access and possible provision of actual resources are distinguished. Since companies do not need to spend large sums of money to purchase these infrastructure equipments, they are in a position to scale up or down their computational power depending on the current demand and thus avoid the situation whereby they are paying for and using more facilities than what is required or, on the other side of the scale, buying little power then being overwhelmed with a high load of processes to complete. This flexibility helps enterprises to overcome the disadvantages of centralized resources such as, the on-premise infrastructure for scale up requires physical acquisitions at additional costs and takes much time.

Cloud platforms are also characterised by high levels of reliability, which is a fundamental ability needed in any contemporary business. With features such as multi-region deployments, load balancing, and the ability to perform automated failover organizations can build systems that are self-adaptive and capable of constantly responding to changes in the environment with the ability to provide uninterrupted service. The one advantage of the cloud is a relative ease and lower cost of implementing disaster recovery measures when using the system to minimize losses in case of a system failure. This means that even if one region is experiencing challenges, the system is always free to transfer to another region in order to continue with the work. In addition, cloud computing has made several cost cutting strategies popular including the use of the pay-as-you-go model. This kind of cost-efficiency is most suitable during organizational dual-busy periods or during the periods of organism's embryonic development. Therefore, most firms are now shifting from those large upfront investments in physical capital because that frees up their budget for better use in finding new opportunities for growth or better technology to develop.

Artificial Intelligence: Enhancing Automation and Decision-Making in Enterprise Systems

AI is a highly significant instrument in the processes of enterprise system change, acting as a means to advance automation and productivity and underpin decisions with data. Therefore, AI allows integration into enterprise architecture and thereby enhances the creation of smart, flexible and high-performance systems capable of managing complex processes and data in real-time. Technologies, including machine learning, natural language processing, and predictive analytics, enable companies to perform repetitive tasks, enhance system functionality and minimize the human factor impact. The use of AI in enterprise systems can be seen as advantageous because it helps serve to perform some tasks that may otherwise be done manually. For instance, the nature of human work can include routine tasks like data inputs, dealing with customer inquiries and system checks – work that can well be done by AI-based automation tools and platforms. Furthermore, their automation optimizes this process while decreasing the possibility of errors as the AI systems perform tasks coherently. It can also have the capability of surveilling systems and equipment in real-time and then diagnosing a problem before it becomes severe and recommending a rectifying action; this will improve system reliability and reduce the duration of system failure. The decision-making capability of AI to choose big data sets and extract meaningful information is also a boon to enterprises for implementing Enterprise Systems. In a big data context, different AI analytical tools can analyse a large amount of data and extract patterns and trends that might not be easily noticed by humans. For instance, predictive analytics enables various enterprises to achieve anticipatory planning instead of reactive problem solutions. Whether this is in terms of demand for the business, supply for products or services, failure of a system or otherwise (Alshadoodee et al., 2022) thus the role of artificial intelligence is evolved in support to decision making. The study takes privates college administration as a varible on which the results rely. The upgrades in innovation have upgraded most techniques for leading business tasks that further develop organizations and administration conveyance. Companies in this area need to wander into digitizing of all industry cycles, business sequences linked to administration and more essential services in educational institutes over time. The need for a proper decision-making support using knowledge management stills create a big gap in the foundation of an effective and efficient eductaional system for the good governance and to improve the image of some institute. The examination interaction has been intended to follow an iterative methodology of information revelation chose for the review. Using the statistical package for social sciences (IBM-SPSS.

DevOps: Accelerating Software Delivery and Enhancing Collaboration in Enterprise Systems

DevOps is now an essential practice of software delivery and infrastructure management in the Fourth Industrial Age as it focuses on enabling collaboration between development and operations with improved and increased speed of software delivering. DevOps involves aligning different practices that help enterprises in accelerating software development and the delivery process and Continuous integration and delivery can be achieved through a combination of different DevOps practices. In addition to this, it fast tracks the release cycle of enterprise applications while at the same time enhancing quality and reliability hence is crucial to organizations that are keen on competing for the ever-shifting market in the current technological world. The major guideline of DevOps is the substitution of manual process with more agile techniques, with emphasis on reducing the work that is done manually in software development, testing and delivery. Forcing integration, deploys, and infrastructure maintenance into automated processes can smoothen dealing and prevent slip-ups that occur from hand scheduling. Both automated testing's and continuous integration help a team guarantee that many problems are found during the development phase so that no tainted code is released. This automation also leads to speedy response time with a view of addressing certain problems, and providing more frequent software releases. DevOps also supports engagement of development, operation and other personnel within an enterprise. In some of the previous styles of software development, the development and operations teams were disparate, and as such, the development team would complete their work only for the operations team to take considerable time to implement them. DevOps is developed in such a way that everyone is accountable for the success as well as the failure of the application development as well as its deployment (Colavita, 2016). This leads to quick problem solving, efficient troubleshooting and enhanced deployment that directly leads to more efficient delivery of the enterprise system. Considering the incorporation of AI/ML based automation tools into the DevOps integrations, enterprises can extend improvements to their software development model. System performance can be continuously monitored with the help of AI tools as well as potential problems and solutions to improve existing performance can be recommended immediately. For instance, AI can estimate which parts of the system could show a bottleneck or failure and can be mitigated once detected by the teams. This predictive capability is valuable in systems that are business-critical for enterprises and where the loss of functionality or loss of efficiency has potentially dire repercussions. That is also beneficial for cloud architectures that are inherently highly scalable and flexible. The cloud platform helps an organization to be able to host and run the applications efficiently in different environments while having built in mechanisms to provide high availability and handling of faults. When adopting DevOps, organizations get advantages of containers, microservices, and other modern practices promoting good scalability, performance, and fault-tolerance in the context of cloud technologies. From the case of cloud computing, scalability and flexibility of deployed applications enable the enterprise to keep abreast of the changing business environment and handle growing workloads without developing bottlenecks. This paper explores why DevOps is crucial for fast delivery of software, better communication and dependable business systems. The processes involved in the development of software are made efficient through key automation and cross-teams work in the DevOps framework. The utilization of AI tools enhances these processes and adds prediction and better automation to them based on the scenario. This is where DevOps amplifies its capabilities when augmented with cloud technologies and enables organizations to extend their applications and deliver high value at scale in a continuously evolving digital environment.

7.3 AI in Decentralized Cloud Systems: Blockchain and Beyond

Distributed computing systems, often referred to as decentralized computing systems, are a collection of discrete systems that use several separate computers working together towards a common goal. In contrast to centralised systems, which have a single server doing all of the work, decentralized systems divide the burden among several nodes. This architecture improves resource use, scalability, and fault tolerance. Different kinds of decentralized systems exist. These consist of distributed databases, blockchain, and peer-to-peer (P2P) networks. For instance, blockchain, which is referred to by the digital currency Bitcoin, is a system in which nodes are network nodes that each have a copy of the ledger. Since it is difficult for a single party to alter the data, this provides transparency, security, and redundancy.

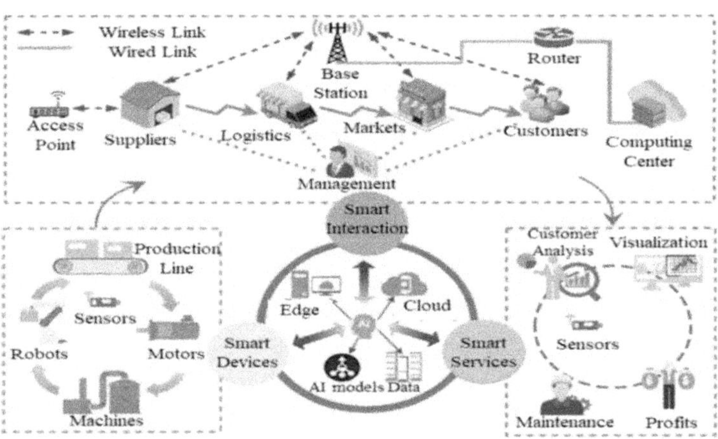

"Smart Industrial Ecosystem: Integration of IoT, AI, and Supply Chain Management"

This image depicts a smart manufacturing ecosystem where interconnected devices, AI, and cloud/edge computing optimize production, supply chains, and business outcomes.
Source: *(Abu-Zaid & Hammad, 2024).*

- **Blockchain**

  The fog, edge, and cloud computing concepts have recently become quite popular in both academia and industry. The security, privacy, and integrity of data in these frameworks have become increasingly important because to their growing use in real-world situations. A threat in this field has been the malicious destruction, theft, and corruption of data brought on by trojans, viruses, and ransomware, among other things. The credibility of the systems depends heavily on maintaining data integrity and making sure that data is not sent by unregistered sources. These systems have very low tolerance since they are employed in mission-critical applications including surveillance, smart cities, transportation, and medical care. It is challenging to provide an ideal plan for data security and integrity maintenance since the majority of Edge devices have computational and storage limits. Blockchain technology has been implemented in the IoT space and other real-time systems to guarantee data security.

  A collection of distributed ledgers called blockchain is capable of recording and monitoring a commodity's worth. A Proof of Work (PoW), which is a hash value that is challenging to generate without altering the PoW of every block that comes before it in the ledger, is produced for each new block of data that is contributed to the system. The Fog system's miners create and verify these Proofs of Work by mining the blocks. A miner publishes a new block in the network after finishing the proof of work, and the rest of the network confirms its legitimacy before appending it to the chain. Furthermore, unless 50% of its distributed copies are individually reformed by performing the identical set of activities, this fraudulent data modification in a blockchain will fail. As a result, changing any data in a blockchain within a strict time limit becomes extremely difficult. Routing, storage, wallet services, and mining are all necessary for network peers to support and function with the blockchain.

There are still a lot of unresolved issues and potential paths for blockchain development in IoT systems. The primary constraint on high-quality data protection and dependability is resource constraints. Such data chains cannot be merged with very complex encryption or key creation due to resource limitations. The use of cryptographic algorithms is restricted. Given the limitations of resources, more effective algorithms can be created. The adaptation of such chains in high fault rate environments, where the edge nodes are susceptible to compromise at any moment, is another crucial avenue. Block revalidation and chain copying from the majority network result in significant network overhead and I/O capacity demands. The majority of frameworks have a single point of failure since they operate in a master-slave paradigm. In diverse surroundings, this is normal. To guarantee redundancy while taking dependability and cost trade-offs into consideration, extensive study is required. Additionally, Fog frameworks are still impacted by blockchain vulnerabilities. It is necessary to create efficient consensus processes that can verify blocks with less block sharing or copying. Readers who are interested may learn more by utilising a comprehensive Blockchain survey.

- **Beyond**

AI is not limited to blockchain and its application is vast in decentralized cloud systems that are rapidly defining the future of distributed computing across various domains. In edge computing, AI is used for analysis and decision-making based on non-critical data to help in IoT, self-driving cars and smart communities. Federated learning enables AI model training without information aggregation, while ensuring data privacy at different locations. New decentralized platforms for buying and selling AI models and computing power help introduce many more people to progressive technologies. Another way that AI helps decentralization is in improving compatibility because when the decentralized systems are heterogeneous and independent, AI allows them to intercommunicate making them portable and expandable. Confidentiality of information in certain applications is maintained by technologies such as homomorphic encryption and secure multiparty computation. Furthermore, AI is at the center of the decentralization of management and operation, with the creation and function of self-running autonomous decentralized organizations (ADOs). Over time, AI helps build quantum algorithms to protect the decentralised systems as quantum computing evolves. Collectively, these developments take decentralized cloud higher and beyond the blockchain, providing a smarter, safer, and more open distributed digital environment,(Kumari et al., 2020).

7.4 Green Cloud Computing: Sustainability Through AI and Automation

The language Grid computing, utility computing, and parallel computing have all contributed to the development of green cloud computing. The current era's advent of Green Cloud Computing is based on traditional resource-sharing technologies.

An ecologically friendly cloud computing system that uses less energy and emits fewer carbon emissions is known as "green cloud computing." Concepts connected to sustainability, such as green ICTs, ecological informatics, environmental informatics, sustainable computing, and green computing, demonstrate the environmentally friendly aspects of ICT products and services. Throughout their existence, ICTs have been researched to support sustainable and green development. By reducing negative consequences that have gotten worse in recent decades, this can significantly improve the environment as it is now. Manufacturers are under tremendous pressure to adhere to environmental regulations and provide goods and services that don't negatively affect the environment (Ficara et al., 2021).

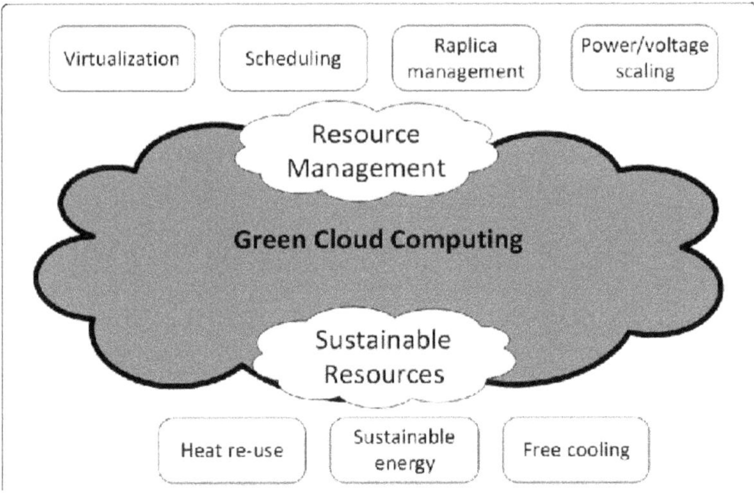

Figure 7.1: illustrates the options for greening cloud computing systems.

Source: *- (Shuja et al., 2017)*

Migration to Sustainable Green Cloud Computing

Green cloud computing is the process of creating, using, and developing digital environments with the primary goal of reducing adverse environmental effects. It entails identifying and creating digital energy-saving techniques to reduce carbon emissions into the environment. It lowers the company costs necessary for operations and conserves energy. Green cloud computing allows users to take use of cloud storage benefits while also reducing adverse climate effects that harm human well-being. Among the various uses of green cloud computing are resource allocation and enhancing the efficiency of communication protocols. Given how quickly cloud computing is developing, more information on it is anticipated to become available. This is becoming an inevitable trend in the construction of green cloud computing data centres. Because they have to oversee an ever-growing number of cloud platform administrators, green cloud data centres have grown in importance. In a green cloud, several processes operate concurrently, requiring high-scope framework resources, which often include numerous servers and cooling offices. Every application is often distributed to many green cloud data centres located in different areas in order to achieve high energy efficiency. For cooling and carrying out various operations, each green cloud data centre typically needs several megawatts of force network energy and eco-friendly power. Furthermore, it is critical to increase the number of servers in the green cloud at its continuous pace due to the rising energy costs of green cloud data centres. The energy consumption of green cloud data centres is rising in tandem with their growing size. Cloud computing is a supercomputer paradigm for collaborative computing that makes advantage of dispersed computing resources, big data centres, high-bandwidth networks, and extensive storage systems. As a result, numerous servers in data centres need to be managed effectively. Green cloud computing is a new computing architecture that is being used to monitor energy use in cloud data centres. Administrators of green cloud computing have given careful consideration to energy productivity and have provided detailed instructions for cutting carbon byproduct emissions in order to save money. In any event, reducing energy use may result in longer time for the service to respond, which impacts service performance; for this reason, finding a balance between energy use and performance is essential. Cloud computing comes in several levels, including utility, grid, and parallel computing. Because data centres are growing so quickly, energy efficiency in green cloud computing has become a conundrum. It can lead to worse performance and slower service response times in terms of both energy and performance. In order to balance the underlying scheduling resources and infrastructure, a group of green communications, cloud computing and innovation, and communications are working to increase data centre computability while lowering carbon dioxide emissions. Resource scheduling is being studied, and there is currently no industry standard for green cloud computing. For cloud computing applications, more communication.

Resources are needed. To meet end-user demands and move massive volumes of data, cloud apps need additional bandwidth. Communication sources that result in loss or congestion due to superfluous switching buffers must be eliminated. The foundation of green computing is the use of optimum algorithms to reduce energy usage. To do this in a green cloud, data centres need to efficiently manage their resources. Processing time and energy consumption are decreased in green cloud computing by allocating jobs to certain resources using optimum scheduling algorithms. Regular tasks arise, and the limited resources in green cloud data centres make it hard to do them all. In green cloud data centres, the admissions controller mechanism is frequently set up to refuse particular tasks and prevent overcrowding. There is no proof that green cloud data centre invoices and job rejections by candidates are related. The jobs are scheduled to operate in green cloud data centres while adhering to their delay constraints using a method called Time-Aware Task Scheduling (TATS), which takes temporal variation into account. By efficiently scheduling all incoming activities from various products to meet task delay-bound restrictions, the Spatial Task Scheduling and Resource Optimisation (STSRO) approach decreased the total service cost. Distributed Energy Resources (DER) for green energy collection can help reduce energy poverty and achieve high network energy efficiency. Although the workload, power consumption model, and experimental setting are not specifically mentioned, the energy and non-service level agreement aware algorithm (EANSA) targets energy reduction by using the environment. Cloud data centers' excessive energy consumption has gained attention in the ICT industry. Using a VANET emulator, the Amazon Elastic Cloud (EC2) computing platform investigates the effectiveness and performance of cloud solutions. By producing ecologically friendly computer hardware, such CPUs with power-saving algorithms, computer manufacturers like Microsoft, Dell, and Hewlett-Packard support green computing. IBM presented criteria to improve its alignment with corporate sustainability policies, examined the components of sustainable ICT, and talked about how it has evolved as a service. In cloud computing settings, soft computing solutions address a variety of work scheduling issues. Appropriate methods for effectively allocating tasks to resources include the genetic algorithm, particle swarm optimisation, ant colony optimisation, and artificial bee colony. In order to optimize resource utilisation in a green cloud environment, we suggest using EC-scheduler, a meta-heuristic bio-inspired technique (Bhattacherjee et al., 2020).

5 Strategies to Migrate Towards Sustainable Green Cloud

In an effort to reduce carbon dioxide emissions and go green, cloud providers are implementing net-zero practices, which allow for changes to IT architecture. In addition to minimizing environmental harm, cloud migration with a net-zero goal will create a sustainable green cloud for the future.

The green cloud goal will be supported by taking into account these tactics, even if not all cloud operations will be energy efficient:

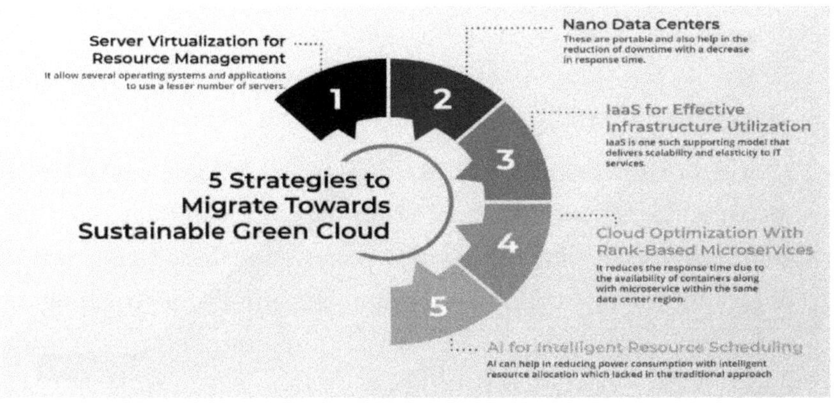

Source: *(Bajpai, 2021)*

1. **Server Virtualization for Resource Management:** The solution to data centers' excessive power usage is virtualization. Using a single resource as several components to make optimal use of

resources (including energy) is the idea underlying virtualization. It is necessary for data centres to provide virtual infrastructures that enable several operating systems and applications to utilize fewer servers. The following are some direct effects of such virtualization on energy efficiency:

- This, virtualised programs may automatically fail over.

- The distribution of resources is coordinated.

- Significant gains in server utilisation indicate a reduction in server requirements.

2. **Nano Data Centers:** These data centres, which are dispersed over the globe, are energy-efficient due to their comparatively lower size and greater proliferation. With a shorter reaction time, these portable micro data centres also aid in lowering downtime. Additionally, they may self-adapt or self-scale and have a high service proximity.

3. **IaaS for Effective Infrastructure Utilization:** The ideal approach to use resources has always been as-a-service with the cloud, and Infrastructure as a Service (IaaS) is one such supporting paradigm that gives IT services scalability and elasticity. The virtual machine (VM) is the primary processing component in data centres. Therefore, it is crucial to lower the energy it uses.

4. **Cloud Optimization with Rank-Based Microservices:** In order to save the environment, businesses that value greener cloud computing are using green cloud optimisation. Due to the flexibility of service delivery, cloud services are increasingly moving away from monolithic architectures and towards microservices, which must be updated for green cloud computing.

Many interdependent microservices operate on cloud nodes, and establishing a rank-based profile for these microservices through container provisioning and distribution across various nodes prior to cloud service execution reduces energy usage. Due to the availability of containers and microservices in the same data centre location, this also speeds up response times.

AI for Intelligent Resource Scheduling: AI can assist in cutting power usage by allocating resources intelligently, something that the conventional method did not do. Resource scheduling plays a crucial role in lowering energy usage since cloud customers' needs vary. Data centres have a major problem with resource scheduling, which AI-enabled technology resolves with intelligent decision-making capabilities.

The idea of green cloud computing is being developed in an effort to preserve our environment. The green cloud can save billions of dollars in energy costs in addition to reducing carbon emissions. It's more about identifying digital energy-saving solutions, such as modifying infrastructure, data storage, deployment, etc., that will lower corporate costs and help cut carbon emissions.(CHRISPIM, 2021).

7.5 Ethical and Societal Implications of AI in the Cloud

Data analytics and decision-making skills have been completely transformed by the incorporation of Artificial Intelligence (AI) into cloud-based Business Intelligence (BI) platforms. Businesses may now more efficiently analyze enormous volumes of data, spot trends, and derive useful insights by utilising AI. However, as the adoption of AI-driven BI solutions grows, so do the ethical challenges associated with their implementation. Issues such as data privacy, algorithmic bias, lack of transparency, and the environmental impact of cloud computing present significant hurdles that businesses must address to maintain trust and compliance in an increasingly digital landscape. Ethical concerns in AI stem largely from the technology's reliance on data, algorithms, and decision-making processes that often operate as opaque systems, making it difficult to ensure accountability and fairness. In the same respect, the expansion of cloud environments is a relatively new area that has new risks associated with data sovereignty, misuse, and security risks. Addressing these ethical issues while celebrating and harnessing theшка innovative value of BI pals important to organizations interested in creating responsible and sustainable BI. Therefore, ethical issues in cloud environment which may occur while implementing the AI based BI systems are discussed in this paper. It analyses ways of managing risks concerned with that data misuse, algorithmic bias, and openness of the decision-making process. Second, the discussion also focuses on the long-term effects of AI on cloud computing in a bid to reduce the adverse ecological impact on our environment. In this case, the key concerns can be solved effectively by the companies in order to create a position of ethical pioneers and defend competitive benefits in the digital environment. By providing a synthesis of these issues in this paper, this work aims at infusing the conversation with a richer consciousness of how and/or where organizations can ethically use or advance AI in cloud based BI frameworks that are relevant while preserving the values of being responsible, inclusive, and sustainable (Ali, 2024).

Ethical Challenges in Cloud Environments

The adaption of AI in cloud-based BI systems has promising future in decision-making systems; however, few ethical issues concern. These potential issues are mainly concerned with data protectionist with algorithms and fairness, and lastly, the issue of explain ability which is a task of achieving both creativity and professionalism.

- **Data Privacy and Security:** Arguably, amongst some of the vital ethical issues ever in a cloud environment is that of security and privacy of data. Given that many businesses organize data both on customers and the business processes in various cloud servers, there are increased dangers of having the data leaked, accessed and misused. For any organization a balancing act must made between legal requirements such as GDPR or CCPA and protection of individual and organizational data. This requires enabling enhanced methods of encryption, users' authorization, and continuous threat surveillance mechanisms. Moreover, today's businesses also require compliance to the data sovereignty laws that dictate the location of data.

- **Algorithmic Bias and Fairness:** The core of predictive analytics in BI systems are based on AI algorithms, but they can be biased. When using the wrong data or incorrect training methods, the algorithms created wind up discriminating between one group and another. For instance, a particular AI model that has been trained from the past hiring data might recommend hiring analytics with gender and race discrimination. To reduce such risks, organisations should consider diverse datasets, periodically review their algorithms, and use technique called fairness-enhancing technologies. This proactive approach make sure that AI system enhances the acknowledgment and use of the principle of equality.

- **Transparency and Accountability:** Many AI models are 'black boxes' making it quite difficult for anyone to understand the manner in which they arrive at a particular decision. Lack of transparency can further be execute from stakeholders to trust AI systems without understanding how conclusion has been arrived. In order to address this, businesses should adopt explainable artificial intelligence or XAI technologies, that offer interpretability alongside the same effectiveness. Furthermore, the formulation of accountability structures guarantees that auditable decisions are made by individuals to maintain the confidence of the stakeholder in the system. By solving these issues, companies will be able to develop AI-based BI systems that make a focus on innovation and responsibility for advancing through digital transformation.

Navigating Ethical Responsibility in Cloud-Based AI Systems

AI as a service is making BI systems flexible to support growing business requirement for data analysis at large scale. But the ethical implications of such advancement come with an imperative to develop practical strategies for solving challenges that come with the advancement in a technology without negating the technology.

1. **Balancing Scalability with Environmental Impact:** Scaling up of the cloud environments allows for large volumes of data processing for use in the prediction models, and other AI applications. However, this scalability most frequently results in negative impacts on the environment. The electric consumption by these computing facilities is unexpectedly high, which has fuelled talks about the environmental impact of big data. In order to do this, businesses must invest in green cloud technologies, including renewable-energy-powered data centers and energy efficient Cloud-AI algorithms. There are solutions such as green cloud computing to reduce all the negative impacts while satisfying the demand for economies of scale.

2. **Ensuring Ethical Use of AI Predictions:** AI integrated BI systems provide forecasts and analysis that direct strategic business decisions including positioning in the market, defining customers segments to target and resource management. However, it is possible to use these predictions in unethical ways, for example to deceive customers, or misuse the data collected. There are also problems of so-called 'decisional blindness' or revealing certain options as unavailable when, in fact, they are merely inconvenient; in this capacity, firms need to develop criteria for properly and ethically implementing the insights derived from the AI. Enhanced report on how predictions are arrived at and how they are then utilised will go a long way in creating trust with the consumers and stakeholders.

3. **Promoting Inclusive AI Development:** Cloud need to include and incorporate the AI systems to adivse systemic bias, and hence the development must encompass diversity. It makes it easier to prevent a given algorithm from being compromised since it involves contributions from different diverse teams hence result in better decision frames. An important question is how developers embrace ethics committees for cooperation and how they are implementing

inclusive data practices to ensure that AI technologies adhere to universally acceptable ethical standards. Attending to these dimensions of ethical responsibility, business benefit from the use of cloud-based AI systems as opportunities for innovation while staying ethical and responsible. It makes certain that the technology enhances the flow of good and the value of trust in the market by achieving a sustainable competitive advantage (Shuja et al., 2017).

7.6 Future Innovations Shaping Cloud and AI Integration

In the future, cloud computing with AI is anticipated to provide incredible prospects to assist automation and real-time data processing capabilities. In the future, the services can expand on sound data security solutions and raise knowledge and appreciation of AI (Akoh Atadoga et al., 2024)prompting an exploration of tools, techniques, and emerging trends to mitigate the environmental impact of software development and operation. This review provides a critical review of current practices and future directions in sustainable software engineering. In recent years, the software industry has recognized the need to address the environmental footprint of software systems, considering factors such as energy consumption, resource utilization, and carbon emissions. Consequently, a plethora of tools and techniques have emerged to support sustainable software development processes. These range from energy-efficient programming languages and frameworks to eco-friendly software architectures and design patterns. Moreover, methodologies such as Green Software Engineering (GSE.

These are some probable future growth areas for cloud computing using AI.

- **Edge Computing Integration:** As edge computing becomes more widely used in application deployments, AI capabilities will be added to handle massive amounts of data near their source. When it comes to use cases that require quick responses, like self-driving cars or smart cities, this capacity would enable fast decision-making and remove delay.

- **Better Personalization and UI:** Cloud platforms' artificial intelligence (AI) can expand natural language comprehension and advanced recommendation algorithms, enabling ever more individualized experiences. This will promote more smooth communication and customized experiences across all services, including healthcare and e-commerce.

- **Evolution of Security:** Cloud computing security will be significantly advanced by AI, which will produce more advanced threat detection and response capabilities that can be included in next-generation systems. In order to guarantee that businesses can foresee new dangers or prevent them before they ever affect their operations, next-generation goods will have more adaptable defenses, self-healing systems, and predictive security.

- **Quantum Computing Incorporation:** The combination of cloud computing, quantum, and artificial intelligence has the potential to revolutionise. AI skills would be greatly enhanced by quantum computing's capacity to tackle hitherto unsolvable problems at previously unthinkable rates, opening up completely new applications for data analysis, optimisation, and issue resolution.

- **Ethical AI and Governance:** There will be a boundary established for the ethical and governance applications of artificial technology as it develops further. As technology advances, models of frameworks and standards for ethical AI use in the cloud will emerge, including those that address privacy, bias, and transparency issues to guarantee efficacy and equality.

Deep learning enables machines to analyze vast volumes of data obtained through data mining using training data already available in the business. AI will be able to handle data like traffic, improving the precision of judgements made in this sector. Artificial intelligence (AI) has improved system security by enabling automatic threat detection through machine learning. As a result, there are now fewer assaults, which benefits users and businesses alike. While interface technologies demonstrate the complexity and difficulties of regulating transdisciplinary sectors, artificial intelligence-based technologies need changes in a number of legal areas. Traditional design, production, and maintenance procedures might be completely transformed by the application of AI technology in the mechanical engineering industry. Engineers can now produce optimised designs more quickly and efficiently thanks to AI-powered design tools, which improves product performance and shortens development cycles. Furthermore, early equipment failure identification is made easier by predictive/forecasting AI in maintenance systems, which reduces maintenance expenses and downtime.

Potential Challenges & Benefits: Although AI in cloud computing has numerous advantages, there are certain restrictions and difficulties that must be resolved. Data security and privacy are among the most crucial issues. For AI systems to learn how to produce relevant responses, they require incredibly enormous volumes of data, therefore protecting sensitive data also becomes crucial. The problem is that security breaches can still occur in cloud environments, and complying with standards like GDPR might create some disruptions in data handling procedures. Furthermore, because all of the data is kept in one location, any one point of failure might affect the entire dataset, which is why data recovery in a separate cloud area is essential.

The potential for prejudice in AI systems might be another drawback. Since AI models are trained using data, systems will reinforce these biases and may provide unfair results through decision-making if the training data distribution is biassed or does not accurately reflect the underlying truth of what would be anticipated in practical use. Additionally, the complexity of integrating AI solutions on cloud infrastructures may make it impossible to capture, which might lead to computational difficulties or a high resource requirement for businesses. Furthermore, AI on cloud computing presents an additional challenge as these systems can only be operated effectively by a skilled crew, which is a major issue for small businesses with limited investment resources.

It makes it clear that using AI cloud computing has a bright future and the potential to change the paradigm of technology by opening up new avenues for innovation and optimisation from the standpoints of performance evaluation. However, this will inevitably raise new ethical concerns.

7.7 Chapter Summary

Adopting cloud-AI technologies is going to be embedded tightly with enterprise processes as the core of activity that is set to redefine the nature of system architectures, decision-making, and longevity. Integrating the principles of cloud computing, such as scalability, fault tolerance, and cost-effectiveness, integrating the principles of artificial intelligence, such as automation, prediction, and data-driven decision-making, can help organisations optimise efficiency, reliability and flexibility. This results in the faster delivery of software and improved system efficiency through integrating DevOps and AI technologies; decentralised and blockchain technologies guarantee transparency innovative security solutions and scalable architectures. The primary goal of green cloud computing is to reduce energy use and adverse environmental consequences, mostly with the use of green energy supplies and scheduling algorithms. AI governance and responsibilisation is an application of a number of ethical issues such as data protection, bias propensity of artificial intelligence algorithms, and transparency. Emerging trends, including Edge computing, Quantum computing and governance of ethical Artificial Intelligence, will continue as key drivers into the future to shift focus towards Cloud-AI systems as essential platforms for sustainable development in a digital global economy.

Multiple Choice Questions (MCQs)

1. **Which of the following topics is NOT covered in the chapter?**

    a. Building resilient and scalable Cloud-AI infrastructures

    b. Green cloud computing and sustainability

    c. Manual optimization of cloud resources

    d. Ethical implications of AI in the cloud

2. **What is the primary focus of "Building Resilient and Scalable Cloud-AI Infrastructures"?**

    a. Optimizing AI for local systems

    b. Designing cloud systems that can handle high demands and adapt to changes

    c. Eliminating the need for AI in cloud infrastructures

    d. Centralizing all cloud operations

3. **"AI in Decentralized Cloud Systems" emphasizes the role of which technology?**

    a. Blockchain

    b. Centralized AI servers

    c. Quantum computing

    d. Manual coding for AI optimization

4. **What is the primary goal of "Green Cloud Computing"?**

    a. Reducing computational errors in AI systems

    b. Increasing sustainability through AI and automation

    c. Developing faster cloud-based processors

    d. Expanding physical cloud data centers

5. **Ethical considerations in cloud-based AI primarily address:**

    a. Ensuring fairness, transparency, and accountability in AI applications

    b. Building faster and more efficient AI systems

    c. Minimizing hardware costs

    d. Expanding the scope of decentralized system

6. **Future innovations shaping Cloud and AI integration are likely to focus on:**

    a. Enhancing collaboration between cloud systems and advanced AI tools

    b. Replacing automation with manual intervention

    c. Limiting cloud-AI usage to specific industries

    d. Reducing the scalability of cloud infrastructures

7. **Blockchain is relevant to the chapter because:**

    a. It supports decentralized AI systems in the cloud

    b. It eliminates the need for cloud infrastructures

    c. It increases manual intervention in cloud management

    d. It focuses on centralizing data processing

8. **Which of the following is a key benefit of green cloud computing?**

    a. Higher profits for cloud service providers

    b. Reduced environmental impact through sustainable practices

    c. Increased manual intervention in data centers

    d. Decreased reliance on AI for optimization

9. What is one societal implication of AI in the cloud?

    a. Reduced need for skilled professionals

    b. Increased data privacy concerns and ethical challenges

    c. Elimination of decentralized systems

    d. Growth of hardware-based AI solutions

10. The chapter summary likely emphasizes:

    a. Future trends, innovations, and ethical challenges in Cloud-AI integration

    b. A deep dive into technical coding for AI

    c. Limiting cloud computing to traditional infrastructures

    d. Replacing AI systems with human-led processes

Answer

1	2	3	4	5	6	7	8	9	10
c	b	a	b	a	a	a	b	b	a

BIBLIOGRAPHY

Abu-Zaid, R., & Hammad, A. (2024). Applications of AI in Decentralized Computing Systems: Harnessing Artificial Intelligence for Enhanced Scalability, Efficiency, and Autonomous Decision-Making in Distributed Architectures. *Journal of Artificial Intelligence & Cloud Computing, 7.*

Akoh Atadoga, Uchenna Joseph Umoga, Oluwaseun Augustine Lottu, & Enoch Oluwademilade Sodiya. (2024). Tools, techniques, and trends in sustainable software engineering: A critical review of current practices and future directions. *World Journal of Advanced Engineering Technology and Sciences.* https://doi.org/10.30574/wjaets.2024.11.1.0051

Ali, S. (2024). *Securing Cloud-Based AI and Machine Learning Models: Privacy and Ethical Concerns. Ml,* 1–12.

Alshadoodee, H. A. A., Mansoor, M. S. G., Kuba, H. K., & Gheni, H. M. (2022). The role of artificial intelligence in enhancing administrative decision support systems by depend on knowledge management. *Bulletin of Electrical Engineering and Informatics.* https://doi.org/10.11591/eei.v11i6.4243

AnalytixLabs. (2022). *Chatbots and Virtual Assistants: Enhancing Customer Engagement in Marketing.* Medium.

Anantrasirichai, N., & Bull, D. (2022). Artificial intelligence in the creative industries: a review. *Artificial Intelligence Review.* https://doi.org/10.1007/s10462-021-10039-7

App Dynamics. (2024). *Now you see it.* App Dynamics.

AWS. (2024). *Amazon CloudWatch*. AWS.

Bajpai, H. (2021). *5 Strategies of Green Cloud Optimization for Sustainable Eco-friendly Cloud Journey*. Techment.

Bajwa, Munir, J., Nori, U., & Williams, A. (2021). Artificial intelligence in healthcare: transforming the practice of medicine. *Future Healthcare Journal*, *8*(2), e188–e194. https://doi.org/10.7861/fhj.2021-0095

Bhattacherjee, S., Das, R., Khatua, S., & Roy, S. (2020). Energy-efficient migration techniques for cloud environment: a step toward green computing. *Journal of Supercomputing*. https://doi.org/10.1007/s11227-019-02801-0

Bibri, S. E., Krogstie, J., Kaboli, A., & Alahi, A. (2024). Smarter eco-cities and their leading-edge artificial intelligence of things solutions for environmental sustainability: A comprehensive systematic review. In *Environmental Science and Ecotechnology*. https://doi.org/10.1016/j.ese.2023.100330

Blessing, M. (2024). *The Role of AI in Real-Time Financial Reporting and Continuous Auditing*.

Borra, P. (2024). An Overview of Cloud Computing and Leading Cloud Service Providers. *INTERNATIONAL JOURNAL OF COMPUTER ENGINEERING & TECHNOLOGY*, *15*, 122–133. https://doi.org/10.17605/OSF.IO/5HQ4M

Boutin, C. (2023). *NIST Drafts Major Update to Its Widely Used Cybersecurity Framework*. NIST Special Publication 1900-202.

Builtin. (2024). *Cloud Computing in Healthcare: How It's Used and 17 Examples*. Builtin.

Cebula, D. (2024). *How to Build Scalable Cloud Infrastructure?* Netguru.

Challenge, M. L. (2024). *Azure Monitor overview*. Microsoft Learn Challenge.

Chowdhury, S., Budhwar, P., Dey, P. K., Joel-Edgar, S., & Abadie, A. (2022). AI-employee collaboration and business performance:

Integrating knowledge-based view, socio-technical systems and organisational socialisation framework. *Journal of Business Research, 144*, 31–49. https://doi.org/10.1016/j.jbusres.2022.01.069

CHRISPIM, M. C. (2021). Resource recovery from wastewater treatment: challenges, opportunities and guidance for planning and implementation. *Frontiers in Neuroscience.*

Colavita, F. (2016). Devops movement of enterprise agile breakdown silos, create collaboration, increase quality, and application speed. *Communications in Computer and Information Science.* https://doi.org/10.1007/978-3-319-27896-4_17

CSA STAR Registry. (2022). *Find a provider with the right level of security and data privacy for your organization.* CSA STAR Registry.

DataDog. (2024). *Modern monitoring & security.* DataDog.

Dynatrace. (2024). *Unified observability and security.* Dynatrace.

Elassy, M., Al-Hattab, M., Takruri, M., & Badawi, S. (2024). Intelligent transportation systems for sustainable smart cities. *Transportation Engineering, 16*, 100252. https://doi.org/10.1016/j.treng.2024.100252

Elastic. (2024). *Meet the search platform that helps you search, solve, and succeed.* Elastic.

Erich, F., Amrit, C., & Daneva, M. (2017). A Qualitative Study of DevOps Usage in Practice. *Journal of Software: Evolution and Process, 00.* https://doi.org/10.1002/smr.1885

Falatah, M. M., & Batarfi, O. A. (2014). Cloud Scalability Considerations. *International Journal of Computer Science & Engineering Survey.* https://doi.org/10.5121/ijcses.2014.5403

Ficara, A., Curreri, F., Cavallaro, L., De Meo, P., Fiumara, G., Bagdasar, O., & Liotta, A. (2021). Social network analysis: the use of graph distances to compare artificial and criminal networks. *Journal of*

Smart Environments and Green Computing. https://doi.org/10.20517/jsegc.2021.08

Gao, R. X., Krüger, J., Merklein, M., Möhring, H.-C., & Váncza, J. (2024). Artificial Intelligence in manufacturing: State of the art, perspectives, and future directions. *CIRP Annals, 73*(2), 723–749. https://doi.org/10.1016/j.cirp.2024.04.101

geeksforgeeks. (2024). *What is Infrastructure as Code (IaC)?* Geeksforgeeks.

Geeksforgeeks. (2023a). *Cloud Deployment Models.* Geeksforgeeks.

Geeksforgeeks. (2023b). *Scalability and Elasticity in Cloud Computing.* Geeksforgeeks.

Geeksforgeeks. (2024). *Evolution of Cloud Computing.* Geeksforgeeks.

Geeksforgeeks. (2025). *Architecture of Cloud Computing.* Geeksforgeeks.

Gill, N. S. (2024). *Essential Tools and Architecture for Cloud Native Security.* XENONSTACK.

Gillis, A. S. (2024). *Definition What is artificial intelligence as a service (AIaaS)?* TechTarget.

Golightly, L., Chang, V., Xu, Q. A., Gao, X., & Liu, B. S. C. (2022). Adoption of cloud computing as innovation in the organization. *International Journal of Engineering Business Management.* https://doi.org/10.1177/18479790221093992

Google Cloud. (2024). *Google Cloud's Observability.* Google Cloud.

Granica. (2024). *6 Cloud cost management best practices.* Granica.

Greeksforgreeks. (2023). *What is CI/CD?* Greeksforgreeks.

Gupta, U., & Sharma, R. (2023). A Study of Cloud Based Solution for Data Analytics in Healthcare. *2023 6th International Conference on Information Systems and Computer Networks, ISCON 2023.* https://doi.org/10.1109/ISCON57294.2023.10112083

Haefner, N., Parida, V., Gassmann, O., & Wincent, J. (2023). Implementing and scaling artificial intelligence: A review, framework, and research agenda. *Technological Forecasting and Social Change*. https://doi.org/10.1016/j.techfore.2023.122878

Haleem, A., Javaid, M., Asim Qadri, M., Pratap Singh, R., & Suman, R. (2022). Artificial intelligence (AI) applications for marketing: A literature-based study. In *International Journal of Intelligent Networks*. https://doi.org/10.1016/j.ijin.2022.08.005

Hashem, I. A. T., Yaqoob, I., Anuar, N. B., Mokhtar, S., Gani, A., & Ullah Khan, S. (2015). The rise of "big data" on cloud computing: Review and open research issues. *Information Systems, 47*(July), 98–115. https://doi.org/10.1016/j.is.2014.07.006

Igbokwe, I. C. (2023). Application of Artificial Intelligence (AI) in Educational Management. *International Journal of Scientific and Research Publications*. https://doi.org/10.29322/ijsrp.13.03.2023.p13536

Impact. (2024). *What Are the 18 CIS Critical Security Controls in Version 8?* IMPACT: International Journal of Research in Humanities, Arts and Literature (IMPACT: IJRHAL).

Indiamart. (2023). *ISO 27001 Information Security Management System Certification Services*. Indiamart.

Ionescu, S. A., & Diaconita, V. (2023). Transforming Financial Decision-Making: The Interplay of AI, Cloud Computing and Advanced Data Management Technologies. *International Journal of Computers, Communications and Control*. https://doi.org/10.15837/ijccc.2023.6.5735

Jadeja, Y., & Modi, K. (2012). Cloud computing - Concepts, architecture and challenges. *2012 International Conference on Computing, Electronics and Electrical Technologies, ICCEET 2012*. https://doi.org/10.1109/ICCEET.2012.6203873

Johnson, S. G., Simon, G., & Aliferis, C. (2024). *Data Preparation, Transforms, Quality, and Management*. https://doi.org/10.1007/978-3-031-39355-6_8

Josyula, V., Orr, M., & Page, G. (2011). *Cloud computing: Automating the virtualized data center.* Cisco Press.

Kavitha, M. S., & Damodharan, P. (2020). Software as a Service in Cloud Computing. *International Journal Of Recent Advances in Engineering & Technology.* https://doi.org/10.46564/ijraet.2020.v08i04.001

Kollolu, R. (2021). History, Deployment and Service Models Towards the Evolution of Cloud Computing. *SSRN Electronic Journal.* https://doi.org/10.2139/ssrn.3917250

Krupa, J., Lin, K., Flechas, M. A., Dinsmore, J., Duarte, J., Harris, P., Hauck, S., Holzman, B., Hsu, S. C., Klijnsma, T., Liu, M., Pedro, K., Rankin, D., Suaysom, N., Trahms, M., & Tran, N. (2021). GPU coprocessors as a service for deep learning inference in high energy physics. *Machine Learning: Science and Technology.* https://doi.org/10.1088/2632-2153/abec21

Kumari, A., Gupta, R., Tanwar, S., & Kumar, N. (2020). Blockchain and AI amalgamation for energy cloud management: Challenges, solutions, and future directions. *Journal of Parallel and Distributed Computing.* https://doi.org/10.1016/j.jpdc.2020.05.004

Labs, G. (2024). *Named a Leader in the Gartner® Magic Quadrant™ for Observability Platforms.* Grafana Labs.

Lins, S., Pandl, K. D., Teigeler, H., Thiebes, S., Bayer, C., & Sunyaev, A. (2021). Artificial Intelligence as a Service. *Business & Information Systems Engineering.* https://doi.org/10.1007/s12599-021-00708-w

Mancino, G. (2022). *The Role of Autonomous Vehicles and AI in Smart Transportation Systems: Enhancing Traffic Management, Route Optimization, and Predictive Maintenance.* https://doi.org/10.13140/RG.2.2.23868.04483

Microsoft. (2021). *What is IaaS? Infrastructure as a Service | Microsoft Azure.* Microsoft.

NIST. (2018). Framework for Improving Critical Infrastructure Cybersecurity [v1.1 Draft]. *National Institute of Standards and Technology.*

Nordlayer. (2023). *Cloud Security architecture & principles.* Nordlayer.

NordLayer. (2022). *Identity and Access Management best practices.* Nordlayer.

Nur Fitria, T. (2021). *Artificial Intelligence (AI) In Education: Using AI Tools for Teaching and Learning Process.*

Ocean, D. (2024). *13 Cloud Monitoring Tools to Ensure Optimal Cloud Performance and Drive Business Success.* Digital Ocean.

Odun-Ayo, I., Ananya, M., Agono, F., & Goddy-Worlu, R. (2018). Cloud Computing Architecture: A Critical Analysis. *Proceedings of the 2018 18th International Conference on Computational Science and Its Applications, ICCSA 2018.* https://doi.org/10.1109/ICCSA.2018.8439638

PagerDuty. (2024). *Transform Your Operations.* PagerDuty.

Pais, S., Cordeiro, J., & Jamil, M. L. (2022). NLP-based platform as a service: a brief review. *Journal of Big Data, 9.* https://doi.org/10.1186/s40537-022-00603-5

Parker, O. (2023). *Cloud Computing and AI in Education: Revolutionizing Personalized Learning While Ensuring Data Security.* https://doi.org/10.13140/RG.2.2.24764.65921

Pastore, S. (2013). The Platform as a Service (PaaS) Cloud Model: Opportunity or Complexity for a Web Developer? *International Journal of Computer Applications.* https://doi.org/10.5120/14225-2435

Patil, D. (2024). *Artificial intelligence in retail and e-commerce: Enhancing customer experience through personalization, predictive analytics, and real-time engagement.*

Prasad, R., & Makesh, D. (2024). *Impact of AI on Media & Entertainment Industry* (pp. 41–71).

Prometheus. (2024). *From metrics to insight*. Prometheus.

Rajak, P., Ganguly, A., Adhikary, S., & Bhattacharya, S. (2023). Internet of Things and smart sensors in agriculture: Scopes and challenges. *Journal of Agriculture and Food Research, 14*, 100776. https://doi.org/10.1016/j.jafr.2023.100776

Rane, J., Mallick, S., Kaya, Ö., & Rane, N. (2024). *Artificial intelligence, machine learning, and deep learning in cloud, edge, and quantum computing: A review of trends, challenges, and future directions* (pp. 1–38). https://doi.org/10.70593/978-81-981271-0-5_1

Rane, N. L., Mallick, S. K., Kaya, Ö., & Rane, J. (n.d.). *Tools and frameworks for machine learning and deep learning : A review. 2024,* 80–95.

Rayhan, A., Kinzler, R., & Rayhan, R. (2023). *NATURAL LANGUAGE PROCESSING: TRANSFORMING HOW MACHINES UNDERSTAND HUMAN LANGUAGE.* https://doi.org/10.13140/RG.2.2.34900.99200

Relic, N. (2024). *Intelligent Observability.* New Relic.

Sharma, D. D., Aggarwal, D. D., & Saxena, D. A. B. (2024). Content Based Recommendation System on Netflix Data. *International Journal of Research In Science & Engineering.* https://doi.org/10.55529/ijrise.42.19.26

Shuja, J., Ahmad, R. W., Gani, A., Abdalla Ahmed, A. I., Siddiqa, A., Nisar, K., Khan, S. U., & Zomaya, A. Y. (2017). Greening emerging IT technologies: techniques and practices. *Journal of Internet Services and Applications.* https://doi.org/10.1186/s13174-017-0060-5

Sinha, D. (2024). *Infrastructure as Code (IaC) in DevOps: The Key to Streamlined DevOps Infrastructure Management.* Techahead.

Sornprom, N. (2024). Role of Cloud Computing & Artificial Intelligence in the Logistics & Supply Chain Industry. *Transactions on Machine Learning and Artificial Intelligence*, *12*, 1–13. https://doi.org/10.14738/tecs.126.17867

Splunk. (2024). *Fortune favors the resilient.* Splunk.

Supriyono, Wibawa, A. P., Suyono, & Kurniawan, F. (2024). Advancements in natural language processing: Implications, challenges, and future directions. *Telematics and Informatics Reports*, *16*, 100173. https://doi.org/10.1016/j.teler.2024.100173

Tecxar. (2022). *What does Continuous Integration and Continuous Delivery (CI/CD) mean in the context of DevOps?* Linkedin.

Torres, P., & George, B. (2023). Disruptive transformation in the transport industry: Autonomous vehicles and transportation-as-a-service. *International Journal of Emerging Trends in Social Sciences*, 28–37. https://doi.org/10.55217/103.v14i1.614

Whaiduzzaman, M., Barros, A., Chanda, M., Barman, S., Sultana, T., Rahman, M. S., Roy, S., & Fidge, C. (2022). A Review of Emerging Technologies for IoT-Based Smart Cities. In *Sensors*. https://doi.org/10.3390/s22239271

Yasar, K. (2024). *Definition load balancing.* Tech Target.

Zhong, R. Y., Xu, X., Klotz, E., & Newman, S. T. (2017). Intelligent Manufacturing in the Context of Industry 4.0: A Review. *Engineering.* https://doi.org/10.1016/J.ENG.2017.05.015

ABOUT THE AUTHOR

Vidyasagar Vangala is an accomplished IT professional, author, and thought leader with a passion for Cloud Technologies, Artificial Intelligence (AI) Integration, and Automation. With a career spanning several years, Vidyasagar has established himself as a trusted expert in driving innovative, scalable, and impactful solutions across various industries.

Currently serving as an IT Project Lead, Vidyasagar specializes in leading large-scale projects involving cloud migration, AI-powered transformations, and DevOps-driven automation frameworks. His hands-on experience with platforms like AWS, Azure, Kubernetes, and Terraform has enabled him to deliver results that align with modern business demands. Vidyasagar's expertise extends to designing and implementing robust workflows, ensuring organizations achieve agility, operational efficiency, and resilience in their digital transformation journeys.

Vidyasagar's professional journey reflects a commitment to solving complex business challenges through technology. As a DevOps Cloud Engineer, he has seamlessly integrated AI-driven solutions into cloud environments, optimizing processes and enhancing decision-making capabilities for his clients. His ability to bridge the gap between emerging technologies and practical applications has earned him recognition as a forward-thinking problem-solver and innovator.

This book represents Vidyasagar's belief in the transformative power of technology and his dedication to helping professionals navigate the ever-evolving digital landscape. His writing blends technical depth with actionable guidance, making it accessible to readers at all skill levels.

When not working on ground-breaking projects or writing, Vidyasagar enjoys exploring emerging trends in AI and cloud computing, constantly seeking ways to contribute to the future of technology.

www.ingramcontent.com/pod-product-compliance
Lightning Source LLC
Chambersburg PA
CBHW040806110726
47973CB00008B/120